Improving Air Force Command and Control Through Enhanced Agile Combat Support Planning, Execution, Monitoring, and Control Processes

Robert S. Tripp, Kristin F. Lynch, John G. Drew, Robert G. DeFeo

Prepared for the United States Air Force

Approved for public release; distribution unlimited

PROJECT AIR FORCE

The research described in this report was sponsored by the United States Air Force under Contract FA7014-06-C-0001. Further information may be obtained from the Strategic Planning Division, Directorate of Plans, Hq USAF.

Library of Congress Cataloging-in-Publication Data

Improving Air Force command and control through enhanced agile combat support planning, execution, monitoring, and control processes / Robert S. Tripp ... [et al.].
    p. cm.
  Includes bibliographical references.
  ISBN 978-0-8330-5309-1 (pbk. : alk. paper)
  1. Command and control systems—United States. 2. United States. Air Force—Operational readiness. 3. Military planning—United States. I. Tripp, Robert S., 1944-

    UB212.I525 2012
    358.4'1330410973—dc23

                                                                    2012005320

Published 2012 by the RAND Corporation
1776 Main Street, P.O. Box 2138, Santa Monica, CA 90407-2138
1200 South Hayes Street, Arlington, VA 22202-5050
4570 Fifth Avenue, Suite 600, Pittsburgh, PA 15213-2665
RAND URL: http://www.rand.org/
To order RAND documents or to obtain additional information, contact
Distribution Services: Telephone: (310) 451-7002;
Fax: (310) 451-6915; Email: order@rand.org

Preface

The Air Force has made significant investments to transform operations and combat support functions to meet the challenges of the current defense environment. Some important operational transformation initiatives deal with strengthening numbered Air Force organizations to improve Air Force warfighting processes and organizations, which, in turn, improves support to combatant commanders (CCDRs). In the combat support area, several major initiatives are included in the Expeditionary Logistics for the 21st Century (eLog21) program. These eLog21 initiatives are broad based and large in scope and cover all aspects of the logistics enterprise, including maintenance, distribution, procurement (sourcing), information, financial, and command and control activities. Because of the complexity and importance of these initiatives, the Air Force senior combat support leadership asked RAND Project AIR FORCE (PAF) to assess the initiatives and to identify any major gaps between and among them.

One significant gap identified in the PAF analysis was the absence of agile combat support (ACS) planning, execution, monitoring, and control, which are sometimes referred to as *agile combat support command and control* (ACS C2) within the Air Force.[1] CCDRs are committing forces to operations without a full understanding of resource

---

[1] ACS planning, execution, monitoring, and control processes (similar to the Air Force monitor, assess, plan, execute [MAPE] model) are an integral part of Air Force enterprise and joint command and control capability. In the revised copy of Air Force Doctrine Document (AFDD) 1, dated November 12, 2010, *ACS C2* is defined as a master capability necessary for Air Force enterprise command and control.

constraint risks. Among other factors, budgetary constraints, the inability to perfectly predict resource demands, the need to shift funding from one category to another to meet unanticipated needs, and the occurrence of unanticipated world events that require intervention all contribute to having imbalances between needed ACS resources and those that are available at any given time to simultaneously meet all requirements for contingency and training operations. Hence, the Secretary of Defense (SecDef) and the Joint Chiefs of Staff (JCS) need to allocate scarce ACS resources among competing demands. Specifically, the Air Force lacks comprehensive doctrine, processes, organizations, training, tools, and systems that enable combat support functions to allocate and utilize limited resources to best achieve operational objectives both in contingency operations and during readiness preparation and training. Enhanced ACS planning, execution, monitoring, and controlling processes are critical elements within the overall Air Force command and control enterprise. They support operational planning, directing, coordinating, and controlling activities as defined in Air Force doctrine (AFDD 2-8, 2007b). This monograph identifies current shortfalls and recent improvements and compares the current state of ACS planning, execution, monitoring, and controlling with the suggested implementation actions designed to address shortfalls identified in the 2002 PAF enterprise command and control operational architecture (OA) (Leftwich et al., 2002).

The research reported here was sponsored by the Director of Transformation, Deputy Chief of Staff for Logistics, Installations, and Mission Support (AF/A4I) and conducted within the Resource Management Program of PAF as part of the "Future Vision for Agile Combat Support: Options for Meeting Emerging Mission Requirements" project.

This monograph will interest CCDRs and their staffs, logisticians, planners, operators, and employers of air and space command and control capabilities throughout the U.S. Department of Defense (DoD), especially those involved with command and control of forces during contingency operations.

This monograph is one of a series of RAND publications that address combat support issues. Related publications include the following:

- *A Strategic Assessment of Component Numbered Air Force (C-NAF) Force Postures*, Kristin F. Lynch, John G. Drew, Amy L. Maletic, Robert S. Tripp, Ricardo Sanchez, William A. Williams, Brent Thomas, and Max Woodworth, 2010, not available to the general public.[2] This monograph outlines processes and a methodology for reviewing the current C-NAF Air Force forces (AFFOR) staff force posture—specifically, requirements as established in Office of the Secretary of Defense (OSD) planning scenarios, opportunities that might be provided by distributed operations, and an analysis of the composition of the workforce with an eye to replacing rated officers and active duty personnel.
- *Global Combat Support Basing: Robust Prepositioning Strategies for Air Force War Reserve Materiel [WRM]*, Ronald G. McGarvey, Robert S. Tripp, Rachel Rue, Thomas Lang, Jerry M. Sollinger, Whitney A. Conner, and Louis Luangkesorn (MG-902-AF), 2010. This monograph identifies alternative approaches to storing combat support materiel that satisfy the requirements of deploying forces in an expeditionary environment that more closely resembles the current DoD planning guidance, while reducing total system costs and increasing robustness.
- *Combat Support Execution Planning and Control: An Assessment of Initial Implementations in Air Force Exercises*, Kristin F. Lynch and William A. Williams (TR-356-AF), 2009. This report evaluates the progress the Air Force has made in implementing the TO-BE ACS OA as observed during operational-level command and control warfighter exercises Terminal Fury 2004 and Austere Challenge 2004 and identifies areas that need to be strength-

---

[2] The C-NAF is the component-level organization the Air Force uses to provide operational-level command and control in order to achieve desired effects across a full range of military operations as defined in U.S. Air Force (2006a). See Appendix C for more information about Air Force and joint command structures.

ened.[3] By monitoring ACS processes, such as how combat support requirements for force package options that were needed to achieve desired operational effects were developed, assessments were made about organizational structure, systems and tools, and training and education.

- *A Strategic Analysis of Air and Space Operations Center Force Posture Options*, Robert S. Tripp, William A. Williams, Kristin F. Lynch, John G. Drew, Dahlia S. Lichter, and Laura H. Baldwin, 2008, not available to the general public. Deriving operational requirements from OSD Defense Planning Scenarios, this monograph evaluates options for the future air and space operations center (AOC) force posture by investigating techniques for right-sizing the AOCs, leveraging distributed operations, enhancing the workforce, and consolidating support. Each option holds the AOC enterprise capability constant and analyzes the resources required to support that capability.

- *Supporting Air and Space Expeditionary Forces [AEFs]: Analysis of CONUS [Continental U.S.] Centralized Intermediate Repair Facilities*, Ronald G. McGarvey, James M. Masters, Louis Luangkesorn, Stephen Sheehy, John G. Drew, Robert Kerchner, Ben D. Van Roo, and Charles Robert Roll Jr. (MG-418-AF), 2008. The Air Force asked RAND to carry out analyses needed to determine alternatives for the use of CONUS centralized intermediate repair facilities (CIRFs) that would provide increased maintenance efficiency (compared with traditional, decentralized structures) without reducing combat support capability. This monograph documents a RAND-developed, optimization-based analytic method for CIRF network design that identifies a range of cost-effective alternatives. It also provides commodity-specific results and recommendations.

- *A Framework for Enhancing Airlift Planning and Execution Capabilities Within the Joint Expeditionary Movement System*, Robert S. Tripp, Kristin F. Lynch, Charles Robert Roll Jr., John G. Drew,

---

[3]   In this monograph, we use *TO-BE* to indicate future conditions and *AS-IS* to indicate current ones.

and Patrick Mills (MG-377-AF), 2006. This monograph examines options for improving the effectiveness and efficiency of intratheater airlift operations within the military joint end-to-end multimodal movement system. Using the strategies-to-tasks framework, this monograph identifies shortfalls and suggests, describes, and evaluates options for implementing improvements in current processes, doctrine, organizations, training, and systems.

- *Supporting Air and Space Expeditionary Forces: Expanded Operational Architecture for Combat Support Planning and Execution Control*, Patrick Mills, Ken Evers, Donna Kinlin, and Robert S. Tripp (MG-316-AF), 2006. This monograph expands and provides more detail on several organizational nodes in our earlier work that outlined concepts for an OA for guiding the development of Air Force combat support execution planning and control needed to enable rapid deployment and employment of the AEF. These combat support planning, execution, and control processes are sometimes referred to as *ACS C2* processes.

- *Supporting Air and Space Expeditionary Forces: Lessons from Operation Iraqi Freedom*, Kristin F. Lynch, John G. Drew, Robert S. Tripp, and Charles Robert Roll Jr. (MG-193-AF), 2005. This monograph describes expeditionary combat support experiences during the war in Iraq and compares these experiences with those associated with Joint Task Force Noble Anvil in Serbia and Operation ENDURING FREEDOM in Afghanistan. This monograph analyzes how combat support performed and how combat support concepts were implemented in Iraq, compares current experiences to determine similarities and unique practices, and indicates how well the combat support framework performed during these contingency operations.

- *Supporting Air and Space Expeditionary Forces: Analysis of Combat Support Basing Options*, Mahyar A. Amouzegar, Robert S. Tripp, Ronald G. McGarvey, Edward W. Chan, and Charles Robert Roll Jr. (MG-261-AF), 2004. This monograph evaluates a set of global forward support location (FSL) basing and transportation options for storing WRM. The authors present an analytical framework that can be used to evaluate alternative FSL options. A central component of the authors' framework is an optimization model that allows a user to select the best mix of land-based and sea-based FSLs for a given set of operational scenarios, thereby reducing costs while supporting a range of contingency operations.

- *Supporting Air and Space Expeditionary Forces: A Methodology for Determining Air Force Deployment Requirements*, Don Snyder and Patrick Mills (MG-176-AF), 2004. This monograph outlines a methodology for determining manpower and equipment deployment requirements. It describes a prototype policy analysis support tool based on this methodology, the Strategic Tool for the Analysis of Required Transportation (START), which generates a list of capability units, called *unit type codes*, that are required to support a user-specified operation. The program also determines movement characteristics. A fully implemented tool based on this prototype should prove to be useful to the Air Force in both deliberate and crisis action planning.

- *Supporting Expeditionary Aerospace Forces: An Operational Architecture for Combat Support Execution Planning and Control*, James A. Leftwich, Robert S. Tripp, Amanda B. Geller, Patrick Mills, Tom LaTourrette, Charles Robert Roll Jr., Cauley von Hoffman, and David Johansen (MR-1536-AF), 2002. This report outlines the framework for evaluating options for combat support execution planning and control. The analysis describes the combat support command and control OA as it is now and as it should be in the future. It also describes the changes that must take place to achieve that future state.

- *Supporting Expeditionary Aerospace Forces: Expanded Analysis of LANTIRN [Low Altitude Navigation and Targeting Infrared for Night] Options*, Amatzia Feinberg, Hyman L. Shulman, Louis W. Miller, and Robert S. Tripp (MR-1225-AF), 2001. This report examines alternatives for meeting LANTIRN support requirements for AEF operations. The authors evaluate investments for new LANTIRN test equipment against several support options, including deploying maintenance capabilities with units, performing maintenance at FSLs, and performing all maintenance at CONUS support hubs for deploying units.

- *Supporting Expeditionary Aerospace Forces: Lessons from the Air War over Serbia*, Amatzia Feinberg, Eric Peltz, James A. Leftwich, Robert S. Tripp, Mahyar A. Amouzegar, Russell Grunch, John G. Drew, Tom LaTourrette, and Charles Robert Roll Jr., 2002, not available to the general public. This report describes how the Air Force's ad hoc implementation of many elements of an expeditionary combat support (ECS) structure to support the air war over Serbia offered opportunities to assess how well these elements actually supported combat operations and what the results imply for the configuration of the Air Force combat support structure. The findings support the efficacy of the emerging ECS structural framework and the associated but still-evolving Air Force support strategies.

- *Supporting Expeditionary Aerospace Forces: A Concept for Evolving to the Agile Combat Support/Mobility System of the Future*, Robert S. Tripp, Lionel A. Galway, Timothy L. Ramey, Mahyar A. Amouzegar, and Eric Peltz (MR-1179-AF), 2000. This report describes a vision for the combat support system of the future based on individual commodity study results.

- *Supporting Expeditionary Aerospace Forces: An Analysis of F-15 Avionics Options*, Eric Peltz, Hyman L. Shulman, Robert S. Tripp, Timothy L. Ramey, and John G. Drew (MR-1174-AF), 2000. This report examines alternatives for meeting F-15 avionics maintenance requirements across a range of likely scenarios. The authors evaluate investments for new F-15 avionics intermediate shop test equipment against several support options, including deploying maintenance capabilities with units, performing maintenance at FSLs, or performing all maintenance at the home station for deploying units.

- *Supporting Expeditionary Aerospace Forces: New Agile Combat Support Postures*, Lionel A. Galway, Robert S. Tripp, Timothy L. Ramey, and John G. Drew (MR-1075-AF), 2000. This report describes how alternative resourcing of forward operating locations (FOLs) can support employment timelines for future AEF operations. It finds that rapid employment for combat requires some prepositioning of resources at FOLs.

- *Supporting Expeditionary Aerospace Forces: An Integrated Strategic Agile Combat Support Planning Framework*, Robert S. Tripp, Lionel A. Galway, Paul Killingsworth, Eric Peltz, Timothy L. Ramey, and John G. Drew (MR-1056-AF), 1999. This report describes an integrated combat support planning framework that can be used to evaluate support options on a continuing basis, particularly as technology, force structure, and threats change.

RAND Project AIR FORCE

RAND Project AIR FORCE (PAF), a division of the RAND Corporation, is the U.S. Air Force's federally funded research and development center for studies and analyses. PAF provides the Air Force with independent analyses of policy alternatives affecting the development, employment, combat readiness, and support of current and future air, space, and cyber forces. Research is conducted in four programs: Force Modernization and Employment; Manpower, Personnel, and Training; Resource Management; and Strategy and Doctrine.

Additional information about PAF is available on our website: http://www.rand.org/paf

Contents

Figures

Tables

Summary

Today's defense environment is particularly challenging for two reasons. First, significant portions of the force are continuously engaged in a variety of operations, ranging from active combat to humanitarian assistance, over wide geographical areas where the needs for force projection are often difficult to predict. Even after operations in Iraq and Afghanistan are concluded, it is likely that the world situation will call for worldwide deployment of U.S. forces to support theater security cooperative efforts (with allies) to shape conditions to avoid contingency operations. Second, there is increasing pressure to operate more efficiently. Although there has always been the need to relate combat support resource requirements to operational objectives, today's environment requires quick combat support actions to tailor deployable support packages and sustainment actions to meet specific operational needs. Furthermore, economic pressures are likely to continue and could result in further reductions in the resources set aside to meet contingency operations. In addition to economic pressures, the inability to perfectly predict resource demands, the need to shift funding from one category to another to meet unanticipated needs, and the occurrence of unanticipated world events that require intervention, among other factors, all contribute to having imbalances between needed ACS resources and those that are available at any given time to simultaneously meet all requirements for contingency and training operations. To best use limited resources in providing combat capability, combat support functional areas must work in an integrated fashion across command and control nodes, predicting combat support needs and responding rapidly to dynamic operational needs, and must allocate scarce resources to where they are most needed. To be successful, combat support decisionmakers need immediate access to a broad range of information with the ability to access detailed data when needed.

In response to this operational environment, the Air Force has invested substantial resources in transforming its operations and combat support functions so that it can meet the needs of the CCDRs more effectively and efficiently. In the combat support arena, the Air Force launched several initiatives, such as eLog21 and Air Force Smart Operations for the 21st Century (AFSO21).[1] The Air Force has invested hundreds of hours of senior-leader time to set the direction, thousands of hours of staff time, and millions of dollars in specific transformation initiatives. In 2008, senior combat support leaders asked PAF to evaluate how well the initiatives align with the future vision for logistics and to identify any gaps between or among them.

A significant gap identified during the 2008 analysis was the lack of ACS planning, execution, monitoring, and control processes to support Air Force operations.[2] ACS planning, execution, monitoring, and control are often referred to as *ACS C2* within the Air Force.[3] Combat support processes are an integral part of Air Force enterprise and joint command and control capability (see Figure S.1). The Air Force lacks the doctrine, processes, organizations, training, and tools that enable combat support to function both effectively and efficiently in the new operational environment.

This monograph describes ACS process gaps in more detail and recommends implementation strategies to facilitate changes needed to improve Air Force command and control through enhanced ACS plan-

---

[1]   eLog21 is an umbrella program comprising many different logistics and supply chain transformational initiatives with an overall goal to improve availability and reduce costs and provide the warfighter with the support he or she needs when it is needed. AFSO21 initiatives are intended to improve the effectiveness and efficiency of overall Air Force operations.

[2]   Similar in construct, the Air Force uses the MAPE model when discussing ACS processes.

[3]   In the revised copy of AFDD 1, dated November 12, 2010, ACS C2 is identified as a master capability necessary for Air Force enterprise command and control.

Figure S.1
The Agile Combat Support Planning, Execution, Monitoring, and Control Structure Works with and in Support of the Air Force and Joint Command and Control Structure

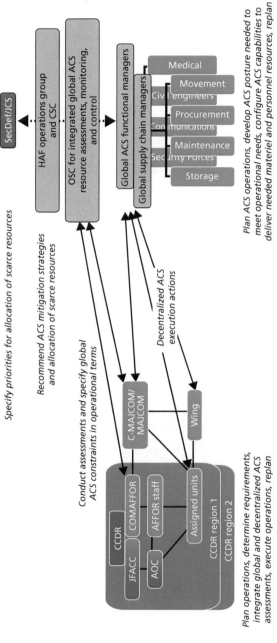

NOTE: This figure depicts a TO-BE vision of Air Force command and control. Joint organizations are shown in purple. Air Force organizations are shown in blue. Gaps where roles and responsibilities have yet to be assigned are shown in orange. This figure is similar to the chart used by the Air Force to describe ACS planning, execution, monitoring, and control processes in Wickman and Battles (2009, slide 3). The difference is that this figure shows the future vision (TO-BE), including the existing organization gaps. HAF = Headquarters Air Force. CSC = combat support center. OSC = operations support center. JFACC = joint force air component commander. COMAFFOR = commander, Air Force forces. C-MAJCOM = component major command. MAJCOM = major command.

RAND MG1070-S.1

ning, execution, monitoring, and control processes. We do not suggest a priority or develop costs associated with implementation plans. The research approach used in this analysis is shown in Figure S.2.

In 2002, RAND researchers developed an OA for combat support. The OA describes ACS planning, execution, monitoring, and control tasks and information flows required to accomplish or support military operations (see Leftwich et al., 2002). We began this work by reviewing the earlier-developed OA from 2002 (and updated in 2006) for adequacy in meeting current and evolving future operational needs. We then identified any process changes resulting from OSD guidance, Air Force guidance, or ongoing transformational initiatives. Finally, we identified any remaining gaps in process, doctrine, training, information systems, and tool sets and proposed options for addressing those gaps.

Figure S.2
Research Approach

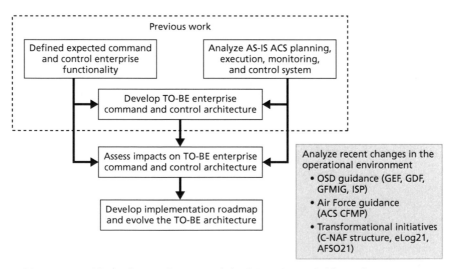

NOTE: GEF = Guidance for Employment of the Force. GDF = Guidance for Development of the Force. GFMIG = Global Force Management Implementation Guidance. ISP = Integrated Security Posture. CFMP = Core Function Master Plan.
RAND MG1070-S.2

What Agile Combat Support Planning, Execution, Monitoring, and Control Shortfalls Exist?

Over the years, incremental improvements have been made to ACS planning, execution, monitoring, and control processes, but some issues persist. ACS process shortfalls span the following five major categories:

- *Poor integration of combat support into operational planning.* Operators often do not involve combat support personnel at the outset of the planning process. Rather, an operational plan is often presented to combat support personnel for them to craft a support plan. But this sequential process can result in infeasible plans or plans that must be altered. Earlier involvement would enable combat support personnel to identify key logistical or operating-location constraints that affect the operational outcomes and allow operators to modify plans to accommodate combat support realities. Operating-location support (for example, civil engineering, security forces, medical) is as critical to successful operations as logistical support that affects mission generation (for example, maintenance, fueling, arming). For instance, site surveys need runway, parking, and infrastructure estimates.
- *Inability to configure combat support processes and resource levels, including supply chain activities, to achieve specific operational objectives; ascertain when ACS process performance or allocated resource levels are not adequate to meet operational objectives; and reconfigure the combat support infrastructure rapidly.* Combat support activities need to be assessed continuously against operational objectives and reconfigured as needed to adapt to changing conditions. This does not happen routinely now. When shortfalls occur, it is difficult to determine why, which makes it difficult to resolve problems. Some functional areas have business rules, tools, and systems to assess system performance. Other areas do not have well-defined, standardized, repeatable processes. Experienced personnel might use ad hoc methods to provide their best estimates. In addition, individual functional analyses are not integrated to give an overall view of combat support capability.

- *Lack of enterprisewide resource assessments to determine whether proposed C-NAF courses of action (COAs) are supportable from a global resource perspective.*[4] When developing proposed COAs, C-NAFs currently assume that global resources will be available to accomplish their assigned missions. They lack visibility into worldwide resource availabilities when developing and executing COAs. They also do not have models and tools or assigned personnel who know how to use available models to access the relevant and authoritative data to identify how constraints in global resource availabilities and process performances might affect operational objectives. As a result, C-NAFs do not know whether their COAs are supportable from a global resource perspective and therefore develop and execute COAs and commit forces to operations with unknown risks.

- *Absence of resource allocation arbitration across competing demands.* There are multiple factors that act together to ensure that, at any given time, there will be differences in available resources and those needed to execute operations. Thus, resource shortages will always exist. To proactively manage allocation of scarce resources across competing areas of responsibility (AORs), allocation priorities need to be developed. Most large operations will likely need to divert resources from units not tasked in the contingency of interest to support units that are tasked. When such reallocations occur, the Air Force currently does not assess how they could affect contingency operations in other AORs, including those from which the resources were reallocated.

- In addition, *ACS shortfalls exist in doctrine, training, and information systems and tools.* Overall, there is a lack of Air Force–wide emphasis on command and control for combat support. ACS objectives, functions, organizational responsibilities, and necessary information flows are not well defined in doctrine. Training

---

[4] The C-NAF is the component-level organization the Air Force uses to provide operational-level command and control in order to achieve desired effects across a full range of military operations as defined in U.S. Air Force (2006a). See Appendix C for more information about Air Force and joint command structures.

opportunities for ACS personnel are also lacking. Exercises and war games do not always accurately address ACS. Training usually focuses on Air Force wing-level skills, not on operational-level skills or how to communicate and operate with other joint services. And, current information and tool shortfalls are consistent with current process shortfalls. The information systems and tool sets that are needed to provide an integrated view of combat support capabilities do not exist today for all functional areas. Some areas have good tools and are able to model combat support capabilities; however, the tools vary across theaters and do not always share information across functional areas.

How Can the Shortfalls Be Eliminated?

We present a vision to address the ACS shortfalls outlined in the previous section. The vision, which has been vetted with Air Force combat support leadership, has three components:

1. *Standardized, repeatable processes to plan, execute, and control combat support activities focused on operationally relevant metrics.* To provide leaders the information they need to make tough trade-off decisions, standardized combat support planning, execution, monitoring, and control processes should be established and defined in doctrine. These processes should draw on capabilities within the Air Force combat support staff (at the C-NAFs and on the Air Staff), global supply chain managers, global ACS functional managers, and a global integration center.

2. *Reliance on the global managers to identify enterprise capabilities and constraints and relay them to C-NAF staffs for use in their contingency planning and execution actions.* There is a need to integrate all the individual supply chain and capability assessments and provide to the C-NAF commander and his or her staff an integrated set of capabilities that can be used in developing COAs to achieve the desired operational effects. Tools are needed for individual functional analyses, as is a method to

integrate individual resources into an overall operational capability, such as mission generation or FOL initial operational capability.

3. *Processes for determining which CCDRs will have priority.* The process using the analysis of global resource shortages (from the first component in this list) to evaluate competing demands and optimize allocation of constrained resources to achieve the desired operational effects should be established and defined in doctrine.

To achieve the ACS vision defined here and close the gaps identified in this analysis, a range of improvements is needed (these recommendations are also summarized in Table S.1). Some are short-term solutions with little implementation cost. Other improvements will take time, resources, planning, and programming.

Table S.1
Steps to Improve Agile Combat Support Planning, Execution, Monitoring, and Control

Goal	Action Needed to Achieve the Goal
Enhance processes	Focus ACS planning, execution, monitoring, and control processes on operational outcomes; identify and separate supply, demand, and integrator processes; include closed-loop feedback and control
Expand doctrine	Delineate roles of ACS nodes, including logistics, operational, and installation staff; Air Force commanders; MAJCOMs; the Air Force Global Logistics Support Center (AFGLSC); and others
Refine training and expand education	Educate Air Force staff officers in ACS planning and staff responsibilities and strategies-to-tasks methodology; assign some promotable "supply-side" officers to "demand-side" organizations and vice versa
Implement systems and tools	Identify critical ACS communications and information-system capabilities needed to assess, monitor, and inform allocation decisions, and update as necessary
Strengthen organizations and instructions	Assign supply, demand, and integrator processes to organizations and functions; modify instructions and other documents to support ACS assessment and control functions

First, ACS supply, demand, and integrator roles need to be clearly defined in doctrine, including what information flows, in what format, and to whom.[5] This could lead to better integration between combat support and operations. In addition, developing a closed-loop planning and execution process, acting within operational decision timelines, with established control parameters against which to track actual combat support performance could aid in making ACS processes more proactive rather than reactive to changing operational requirements.[6] This too could lead to better coordination, timeliness, and accuracy of combat support planning and added value of ACS to the operational community.

The absence of well-defined supply, demand, and integrator processes, delineated in policy, contributes to a shortfall in training. Many ACS personnel do not understand how to apply the nonmarket, resource-constrained strategies-to-tasks and closed-loop frameworks to maximize efficiency and effectiveness.[7] More training and expanded educational opportunities are needed on relating combat support options to the CCDR's campaign plan to achieve joint operational effects.

Decision-support tools and job-performance aids should complement formal courses and exercises. Existing Air Force systems and prototype tools can be leveraged to provide enhanced information and data for ACS planners; however, new tools might need to be developed to provide an integrated view of combat support resource allocations and process performance. Properly integrated information could greatly reduce the risk of operational failure and the need to revise plans midstream, allow a faster transition to operations and better-informed decisions, and facilitate adjustments when necessary.

And finally, global management and control of combat support capabilities could facilitate resource allocation assessments across competing CCDRs to inform tough capability trade-off decisions. These assessments should be used to inform program objective memorandum

---

[5]   See Chapter Four for discussion of supply, demand, and integrator functions.

[6]   See Chapter Four for discussion of closed-loop systems.

[7]   See Appendix A for discussion of the strategies-to-tasks decision framework.

and other budgeting and program decisions. However, with global management comes some risk of single-point failure. Methods to provide continuity of operations (COOP) and to minimize network vulnerabilities need to be developed.

In this analysis, we define processes that are not currently assigned to an organization (shown in orange in Figure S.1). Combat support resource assessment and allocation management could be assigned to permanent organizational nodes dedicated to resource monitoring, prioritization, and reallocation. Additionally, having a standing integration function for combat support resource management could facilitate the incorporation of relevant data into capability assessments and raise the visibility of these assessments in the eyes of the operational community. Air Force senior leadership needs to decide how ACS planning, execution, monitoring, and control nodes might be best organized to carry out their command and control functions. Regardless of the organizational structure adopted, the roles and responsibilities of each ACS organizational node, as well as their interaction with joint combat support nodes, should be clearly defined and documented in Air Force doctrine and guidance, including information needed, processes, and information produced at each node.

Although Air Force transformational initiatives (both operational and in combat support) have moved the Air Force forward in achieving the ACS contingency planning, execution, monitoring, and control TO-BE vision, much remains to be done. This monograph highlights the top-level process, doctrine, policy, training, and organizational changes that need to take place. Using this high-level document, PAF worked with the Air Force to perform a comprehensive review of all the combat support functional capabilities, as identified in the ACS CFMP, and updated the enterprise command and control OA. A follow-on publication updates the details found in the 2002/2006 enterprise command and control OA to reflect the current operational environment. That work focuses on nodal roles and responsibilities. Another follow-on publication provides an incremental approach of how enhanced ACS processes can be incorporated within the Air Force command and control enterprise.

Acknowledgments

Numerous persons inside and outside of the Air Force provided valuable assistance to and support of our work. They are listed here with their rank and position as of the time of this research. We thank Lt Gen Terry Gabreski, Air Force Materiel Command (AFMC) Vice Commander (AFMC/CV) at the beginning of this research; Lt Gen Janet Wolfenbarger, AFMC/CV at the completion of this research; and Lt Gen Loren Reno, Deputy Chief of Staff for Logistics, Installations, and Mission Support (AF/A4/7), for sponsoring this work. We also thank the senior Air Force mentors Lt Gen (R) Joseph Hurd and Lt Gen (R) Michael Zettler for their time and support.

This work would not have been possible without the support of many individuals. At the Air Staff, we thank Patricia Young, AF/A4/7; Grover Dunn, AF/A4I; Maj Gen Robert McMahon and Sue Lumpkins, Deputy Chief of Staff, Logistics, Installations, and Mission Support, Directorate of Logistics; Brig Gen Arthur Cameron III, Deputy Chief of Staff, Logistics, Installations, and Mission Support, Directorate of Resource Integration; Maj Gen Duane Jones, Col Stephen Shea, Lt Col Daryl Cunningham, and Laine Krat, Deputy Chief of Staff, Logistics, Installations, and Mission Support, Directorate of Global Combat Support; and Russell Frasz, Deputy Chief of Staff, Plans and Requirements, Directorate of Operational Planning, Policy and Strategy. In the Directorate of Operations for the Deputy Chief of Staff, Operations, Plans, and Requirements, we thank Col Steven Ruehl, Lt Col James Sturim, Lt Col John Diercks, Lt Col Kari Smith, Allen Wickman, Lt Col Elisabeth Auld, Maj Alexis Kimber, and Lt Col Walter Manwill.

Also at the Air Staff, we thank Chief Scott Heisterkamp, John Ray, Michael Robertson, Col Sid Banks, Col Patricia Battles, Dick Olson, Freddie McSears, William (Dave) Sweet, Nick Reybrock, Col Donald Gleason, and Col Robert Mendenhall.

At AFMC, we thank Lt Gen Thomas Owen, Air Force Materiel Command, Directorate of Logistics (former AFMC/A4), at the completion of this research; Maj Gen Kathleen Close, AFMC/A4 at the completion of this research; Maj Gen Marshall (Keye) Sabol, AFMC, Directorate of Strategic Plans, Programs and Analyses (AFMC/A8/9); Richard Moore, Bob McCormick, Tom Stafford, Col Chris Froehlich, William Santiago, and Molly Waters. From the AFGLSC, we thank Maj Gen Gary McCoy, commander, AFGLSC; Lorna Estep; Col Roger Thrasher, vice commander, AFGLSC; Richard Reed; Michael Howenstine; Lt Col Kevin Gaudette; Mike Niklas; Frank Washburn; Debra Garves; James Weeks; and Lynne Grile.

In the Pacific AOR, we thank Lt Gen Herbert Carlisle, commander, 13th Air Force; Col Creig Rice, 13th Air Force Chief of Staff; Col Gregory Cain, 13th Air Force Chief of Staff; Col Barry White, 13th Air Force Director of Logistics; Col Darlene Sanders, 13th Air Force Director of Logistics; and Elaine Ayers and Capt James Arnett. At Pacific Air Forces, we thank Brig Gen Brent Baker, Pacific Air Forces Director of Logistics, and Russell Grunch.

At Air Forces Central (AFCENT), we thank Brig Gen Richard Shook and the AFCENT staff. At Air Combat Command (ACC), we thank Maj Gen Judith Fedder, ACC Director of Logistics; Col Walter Fulda; Melvin Miles; and Scott Carter. At the Global Cyber Integration Center, we thank Stan Newberry, director; Col John Rudolph; and Desiree Stone. On the Secretary of Air Force (SECAF) staff, we thank Col David Geuting and Deborah Dewitt. At the Joint Staff, we thank Susan McDonald. At the Vehicle and Equipment Management Support Office, we thank SMSgt Scotty Browning. And finally, at the Ogden Air Logistics Center, we thank Mark Johnson, Col Perry Oaks, Mark Brown, Wendy Kierpiec, and the entire Global Ammunition Control Point staff for their time.

At RAND, we are grateful for the support given by Laura Baldwin, Amy Maletic, James Masters, Don Snyder, Jerry Sollinger, and Jin Woo Yi. Their comments and reviews of this work have helped enhance our effort. We would especially like to thank Irv Blickstein and John Halliday for their thorough review of this monograph. Their reviews helped shape it into its final, improved form.

Responsibility for the content of the monograph, analyses, and conclusions lies solely with the authors.

Abbreviations

A1	manpower and personnel directorate
A2	intelligence directorate
A3	air, space, and information operations directorate
A4	logistics directorate
A7	installations and mission support directorate
A8	strategic plans and programs directorate
A9	analyses, assessments, and lessons learned directorate
AALPS	Automated Air Load Planning System
ACC	Air Combat Command
ACS	agile combat support
ACS C2	agile combat support command and control
ADCON	administrative control
AEF	air and space expeditionary force
AEFC	Aerospace Expeditionary Force Center
AEFC/CC	commander, Aerospace Expeditionary Force Center

AEFSG	Air and Space Expeditionary Force Steering Group
AEFTF	Air and Space Expeditionary Force Task Force
AETF	air and space expeditionary task force
AF/A3/5	Deputy Chief of Staff, Operations, Plans, and Requirements
AF/A3O	Deputy Chief of Staff, Directorate of Operations
AF/A3OO	Air Force Operations Group
AF/A4/7	Deputy Chief of Staff, Logistics, Installations, and Mission Support
AF/A4/7P	Deputy Chief of Staff, Logistics, Installations, and Mission Support, Directorate of Resource Integration
AF/A4/7Z	Deputy Chief of Staff, Logistics, Installations, and Mission Support, Directorate of Global Combat Support
AF/A4I	Deputy Chief of Staff, Logistics, Installations, and Mission Support, Directorate of Transformation
AF/A4L	Deputy Chief of Staff, Logistics, Installations, and Mission Support, Directorate of Logistics
AF/A7C	Deputy Chief of Staff, Logistics, Installations, and Mission Support, Directorate of Civil Engineering
AFAFRICA	Air Forces Africa
AFB	Air Force base

AFC2IC	Air Force Command and Control Integration Center
AFCENT	Air Forces Central
AFCESA	Air Force Civil Engineering Support Agency
AFCS	Air Force corporate structure
AFDD	Air Force doctrine document
AFEUR	Air Forces Europe
AFFOR	Air Force forces
AFGLSC	Air Force Global Logistics Support Center
AFI	Air Force instruction
AFIT	Air Force Institute of Technology
AFKOR	Air Forces Korea
AFMC	Air Force Materiel Command
AFMC/A4	Air Force Materiel Command, Directorate of Logistics
AFMC/A8/A9	Air Force Materiel Command, Directorate of Strategic Plans, Programs and Analyses
AFMC/CV	vice commander, Air Force Materiel Command
AFNORTH	Air Forces Northern
AFOG	Air Force Operations Group
AFPAC	Air Forces Pacific
AFPC	Air Force Personnel Center
AFPC/DPW	Air Force Personnel Center, Directorate of Air and Space Expeditionary Force Operations

AFPD	Air Force policy directive
AFRC	Air Force Reserve Command
AFSO21	Air Force Smart Operations for the 21st Century
AFSOC	Air Force Special Operations Command
AFSOF	Air Force special operations forces
AFSOUTH	Air Forces Southern
AFSPC	Air Force Space Command
AFSTRAT-GS	Air Forces Strategic–Global Strike
AFSTRAT-SP	Air Forces Strategic–Space
AFTRANS	Air Forces Transportation
AFUTL	Air Force Universal Task List
ALC	air logistics center
ALEX	Agile Logistics Evaluation eXperiment
AMC	Air Mobility Command
AMETLS	agency mission essential task list
ANG	Air National Guard
AOC	air and space operations center
AOR	area of responsibility
APPG	Annual Planning and Programming Guidance
ART	Air Expeditionary Force Reporting Tool
ATO	air tasking order
BEAR	basic expeditionary airfield resources

BOS	base operating support
C2	command and control
CAF	combat air forces
CAOC	combined air and space operations center
CBP	capabilities-based planning
CCDR	combatant commander
CCRCT	contingency centralized repair center time
CCRF	contingency centralized repair facility
CDDOC	U.S. Central Command Deployment and Distribution Operations Center
CFACC	combined force air component commander
CFMP	Core Function Master Plan
CIRF	centralized intermediate repair facility
CJCS	Chairman of the Joint Chiefs of Staff
CJCSG	Chairman of the Joint Chiefs of Staff guide
CJCSI	Chairman of the Joint Chiefs of Staff instruction
CJCSM	Chairman of the Joint Chiefs of Staff manual
CJCSN	Chairman of the Joint Chiefs of Staff notice
C/JFACC	combined or joint force air component commander
C-MAJCOM	component major command
C-MAJCOM/CC	commander, component major command
CMOS	cargo movement operations system
C-NAF	component numbered Air Force

C-NAF/CC	commander, component numbered Air Force
COA	course of action
COCOM	combatant command
COMACC	commander, Air Combat Command
COMAFFOR	commander, Air Force forces
CONOPS	concept of operations
CONPLAN	contingency plan
CONUS	continental United States
COOP	continuity of operations
CRF	centralized repair facility
CSA	combat support agency
CSAF	Chief of Staff, U.S. Air Force
CSC	combat support center
CST	combat support team
CWT	customer wait time
DCAPES	Deliberate and Crisis Action Planning and Execution Segments
DDOC	deployment and distribution operations center
DEPORD	deployment order
DLA	Defense Logistics Agency
DOC	designed operational capability
DoD	U.S. Department of Defense
DODD	U.S. Department of Defense directive
DODI	U.S. Department of Defense instruction

DOTMLPF	doctrine, organization, training, materiel, leadership and education, personnel, and facilities
DPDRT	Deployment Processing Discrepancy Reporting Tool
DRRS	Defense Readiness Reporting System
DRU	direct reporting unit
ECS	expeditionary combat support
EDDOC	U.S. European Command Deployment and Distribution Operations Center
eLog21	Expeditionary Logistics for the 21st Century
ESORTS	Enhanced Status of Resources and Training System
EXORD	execute order
FAM	functional area manager
FOA	field operating agency
FOC	full operational capability
FOL	forward operating location
FSL	forward support location
FY	fiscal year
GACP	Global Ammunition Control Point
GDF	Guidance for Development of the Force
GEF	Guidance for Employment of the Force
GFM	Global Force Management
GFMIG	Global Force Management Implementation Guidance

GIC	global integration center
GSORTS	Global Status of Resources and Training System
HAF	Headquarters Air Force
ICIS	Integrated Consumable Item Support
ICP	inventory control point
IDO	installation deployment officer
IDS	Integrated Deployment System
IOC	initial operational capability
ISP	Integrated Security Posture
ISR	intelligence, surveillance, and reconnaissance
IT	information technology
J-3/5	Joint Staff, Operations/Strategic Plans and Policy Directorate
J-4	Joint Staff, Logistics Directorate
J-4 RD	Joint Staff, Logistics Directorate, Readiness Division
J/AMETL	joint and agency mission essential task list
JCS	Joint Chiefs of Staff
JDDOC	joint deployment and distribution operations center
JFACC	joint force air component commander
JFC	joint force commander
JFP	joint force provider
JMETL	joint mission essential task list

JOPES	Joint Operation Planning and Execution System
JOPP	joint operation planning process
JP	joint publication
JPG	joint planning group
JQRR	joint quarterly readiness review
JRS	Joint Reporting Structure
JRSOI	joint reception, staging, onward movement, and integration
JSCP	Joint Strategic Capabilities Plan
JSPS	Joint Strategic Planning System
JTF	joint task force
JTF NA	Joint Task Force Noble Anvil
JTIMS	Joint Training Information Management System
JTS	Joint Training System
LANTIRN	low altitude navigation and targeting infrared for night
LOGMOD	logistics module
LRC	logistics readiness center
LRU	line replaceable unit
LSA	logistics sustainability analysis
LSC	logistics support center
MAF	mobility air forces
MAJCOM	major command

MAJCOM/CC	commander, major command
MAPE	monitor, assess, plan, execute
MCO	major combat operation
MDS	mission design series
MET	mission essential task
METL	mission essential task list
MOE	measure of effectiveness
NAF	numbered Air Force
NCA	national command authorities
NGB	National Guard Bureau
NMS	National Military Strategy
NRCT	network repair center time
OA	operational architecture
OCONUS	outside the continental United States
OEF	Operation ENDURING FREEDOM
OIF	Operation IRAQI FREEDOM
ONA	Operation NOBLE ANVIL
OODA	observe, orient, decide, act
OPCON	operational control
OPLAN	operations plan
OPORD	operations order
OPR	office of primary responsibility
OSC	operations support center
OSD	Office of the Security of Defense

OSF	operational support facility
PAD	program action directive
PAF	RAND Project AIR FORCE
PDDOC	U.S. Pacific Command Deployment and Distribution Operations Center
PDDOC-K	U.S. Pacific Command Deployment and Distribution Operations Center—Korea
POC	point of contact
POL	petroleum, oils, and lubricants
POM	program objective memorandum
PPBE	Planning, Programming, Budgeting, and Execution
RC	Reserve Component
RFC	request for capabilities
RFF	request for forces
ROMO	range of military operations
RSP	readiness spares package
SAF/XC	Secretary of the Air Force, Office of Warfighting Integration and Chief Information Officer
SCF	service core function
SE	support equipment
SECAF	Secretary of the Air Force
SecDef	Secretary of Defense
SIPT	scheduling integrated process team

SORTS	Status of Resources and Training System
SPG	Strategic Planning Guidance
SRU	shop replaceable unit
START	Strategic Tool for the Analysis of Required Transportation
TACON	tactical control
TDS	theater distribution system
TMO	traffic management office
TPFDD	time phased force and deployment data
TTP	tactics, techniques, and procedures
UDM	unit deployment manager
UHTS	unsourced or hard to source
UJTL	Universal Joint Task List
UNAAF	Unified Action Armed Forces
USAFE	U.S. Air Forces Europe
USCENTCOM	U.S. Central Command
USEUCOM	U.S. European Command
USFK	U.S. Forces Korea
USJFCOM	U.S. Joint Forces Command
USNORTHCOM	U.S. Northern Command
USPACOM	U.S. Pacific Command
USSOCOM	U.S. Special Operations Command
USSOUTHCOM	U.S. Southern Command
USSTRATCOM	U.S. Strategic Command

USTRANSCOM	U.S. Transportation Command
UTC	unit type code
WRM	war reserve materiel

CHAPTER ONE

Introduction, Background, and Motivation

Command and control (C2) of air and space power is a fundamental function of the Air Force that enables the United States to conduct operations that accomplish specific military objectives. According to Air Force Doctrine Document (AFDD) 1, command and control is defined as

> the exercise of authority and direction by a properly designated commander over assigned and attached forces in the accomplishment of the mission. C2 includes both the process by which the commander decides what action is to be taken and the systems that facilitate planning, execution, and monitoring of those actions. Specifically, C2 includes the battlespace management process of planning, directing, coordinating, and controlling forces and operations.
>
> C2 involves the integration of a system of procedures, organizational structures, personnel, equipment, facilities, information, and communications designed to enable a commander to exercise authority and direction across the range of military operations. Air and space forces conduct the C2 function to meet strategic, operational, and tactical objectives. (AFDD 1, 2003, pp. 49–50)[1]

---

[1] The updated draft version of AFDD 1, currently in top line coordination, states,

Command and control is the exercise of authority and direction by a properly designated commander over assigned and attached forces in the accomplishment of the mission. Command and control functions are performed through an arrangement of personnel, equipment, communications, facilities, and procedures employed by a commander in

Another fundamental function of the Air Force is agile combat support (ACS), defined in AFDD 1 as

> the timely concentration, employment, and sustainment of US military power anywhere. . . . [It] creates, sustains, and protects all air and space capabilities to accomplish mission objectives across the spectrum of military operations. (AFDD 1, 2003, p. 48)

Prior RAND Project AIR FORCE (PAF) research found that ACS planning, execution, monitoring, and control processes (often referred to as *ACS C2* processes within the Air Force) are not adequately defined and delineated in doctrine, and tools or systems to support these ACS processes are lacking (Leftwich et al., 2002; Mills et al., 2006).[2] Although sometimes compared with commercial supply chains, a military ACS system is very different from commercial supply chain operations. Military supply chain management faces conditions that are unlike most, if not all, commercial supply chain operations. Commercial supply chains do not pick up their production plants, move them thousands of miles away, operate in what could be a hostile or austere environment, and expect their production systems to operate within hours after arrival.[3] The purpose of this analysis is to review current ACS processes, identify and describe where shortfalls or major gaps exist, and suggest mitigation strategies to facilitate needed changes for an efficient and effective global combat support system.

---

planning, directing, coordinating, and controlling forces and operations in the accomplishment of the mission (JP [Joint Publication] 1-02). This core function deals with the capabilities necessary to support theater air C2, space C2, cyberspace C2, nuclear C2, air mobility C2, and agile combat support C2. Capabilities in this function integrate the strategic, operational, and tactical levels of war across the ROMO [range of military operations].

[2] ACS planning, execution, monitoring, and control (similar to the Air Force monitor, assess, plan, execute [MAPE] model) to support Air Force operations (contingency, readiness preparation, or training) is often referred to within the Air Force community as *ACS C2*. Combat support processes are an integral part of Air Force enterprise and joint command and control capability.

[3] Thus, commercial supply chain best practices do not always apply to military ACS processes.

Background and Research Motivation

Today's defense environment is particularly challenging for two reasons. First, significant portions of the force are continuously engaged in a variety of operations, ranging from active combat to humanitarian assistance, over wide geographical areas where the needs for force projection are often difficult to predict. Even after operations in Iraq and Afghanistan are concluded, it is likely that the world situation will call for worldwide deployment of U.S. forces to support theater security cooperative efforts (with allies) to shape conditions to avoid contingency operations. Second, there is increasing pressure to operate more efficiently. And, although there has always been the need to relate combat support resource requirements to operational objectives, today's environment requires quick combat support actions to tailor deployable support packages and sustainment actions to meet specific operational needs. Furthermore, economic pressures are likely to continue and could result in further reductions in resources that are set aside to meet contingency operations. In addition to economic pressures, the inability to perfectly predict resource demands, the need to shift funding from one category to another to meet unanticipated needs, and the occurrence of unanticipated world events that require intervention, among other factors, all contribute to having imbalances between needed ACS resources and those that are available at any given time to simultaneously meet all requirements for contingency and training operations. As a consequence, combat support functional areas must work in an integrated fashion across command and control nodes, providing predictions of combat support needs and rapid responses to dynamic operational needs, and allocate scarce resources to where they are most needed. To be successful, combat support decisionmakers need immediate access to a broad range of information with the ability to access detailed data when needed. ACS is broader than just logistics (the A4 [logistics directorate]). It includes personnel (the A1), services (also in the A1), communications (the A6), and installations and mission support (the A7).[4] In this analysis, we consider all aspects of combat support and how integrated ACS can help the warfighter achieve the desired operational effects.

Today, in most cases, ACS planning, execution, monitoring, and control processes are ad hoc, with only a few functional areas managing capabilities and resources from an enterprise perspective. Munitions, for example, has a global requirements determination process and an allocation board to distribute assets worldwide. Fuels and civil engineering capabilities are also managed globally. Other functional-area processes might not be as well defined or standardized—for example, services personnel and equipment and vehicles are not managed globally. They are managed theater by theater without an enterprise view of worldwide capability. However, combat support of military operations remains successful primarily because of the efforts of individuals in the combat support community who overcome difficulties in current (AS-IS) processes, systems, tools, and training. Since the Air Force will continue to operate in a resource-constrained environment in the future, ACS should support trade-off and allocation decisionmaking with standardized, analytic processes.

Transformation

The Air Force and the U.S. Department of Defense (DoD) recognize the need to transform themselves to meet existing and emerging global requirements with limited resources. The Air Force has made significant investments to transform operations and combat support functions to improve its capabilities to meet the challenges posed by the current defense environment. For example, Program Action Directive (PAD) 06-09 established component numbered air forces (C-NAFs) as the Air Force component organizational structure to enhance operational-level command and control of air, space, and information operations across a broad range of engagements (PAD 06-09, 2006b; U.S. Air Force, 2006a). The air and space operations center (AOC), a part of the C-NAF, was designated as a weapon system whose process-oriented

---

[4]   See Appendix C for a description of the Air Force forces (AFFOR) staff and the functional directorates (A1 through A9).

focus is on producing war plans and executing them to achieve strategic and tactical objectives.[5]

The Air Force has also begun to transform its logistics enterprise so that it is both more responsive in meeting combatant commander (CCDR) needs and more efficient in training, organizing, and equipping forces for operational missions. In response to needed changes to the logistics enterprise, senior logistics leaders launched an Expeditionary Logistics for the 21st Century (eLog21) program of initiatives to modernize and streamline its logistics operations to address the challenges of this more demanding environment within limited budgets.[6] The goal of eLog21 was both to make the logistics enterprise more responsive and effective in meeting changing operational demands and to reduce operating and sustainment costs. Simultaneously, others in the Air Force and DoD introduced additional initiatives intended to improve the effectiveness and efficiency of overall Air Force operations. These programs, such as Air Force Smart Operations for the 21st Century (AFSO21), deal with related, and sometimes overlapping, objectives.

The scope of the Air Force logistics transformation is large and affects the processes, doctrine, personnel, facilities, equipment, information, and tool sets that encompass all Air Force supply chain operations, including flightline, depot, original equipment manufacturer, contract, and other suppliers of services and materiel. The Air Force has expended considerable effort improving demand forecasting and accurately estimating safety stock levels (or buffers) to mitigate potential constraints. Other efforts have focused on reducing the transportation time between the user and the repair site. Although these are worthy goals, they are often evaluated independently; improvements within

---

[5]   The C-NAF is the component-level organization the Air Force uses to provide operational-level command and control in order to achieve desired effects across a full range of military operations as defined in U.S. Air Force (2006a). See Appendix C for more information about Air Force and joint command structures.

[6]   eLog21 is an umbrella program comprising many different logistics and supply chain transformational initiatives with an overall goal of improving availability, reducing costs, and providing the warfighter with the support he or she needs when it is needed.

one stovepipe might optimize that individual component with little (or perhaps even negative) impact on enterprise performance.

In 2008, senior logisticians in the Air Force asked PAF to examine the logistics transformational initiatives to assess how well they align with the future vision of the logistics enterprise and to identify any gaps between and among the initiatives that might prevent achieving the capabilities contained in the vision. One gap identified during this analysis was the need for clear definition of ACS processes to meet specific operational contingency, readiness preparation, and training requirements. These ACS processes work within the Air Force command and control enterprise and must perform within the operational planning observe, orient, decide, act (OODA) loop. As a result of this gap analysis, in March 2009, the Deputy Chief of Staff for Logistics, Installations, and Mission Support (AF/A4/7) and the Vice Commander of Air Force Materiel Command (AFMC) asked PAF to accomplish the following tasks:

- Review the RAND-developed enterprise command and control operational architecture (OA) from 2002 (updated in 2006) for adequacy of meeting current and evolving needs given progress made in eLog21 and C-NAF transformation efforts.
- Identify any process changes that result from these transformation efforts.
- Highlight remaining gaps in the TO-BE OA.
- Update ACS processes, doctrine, training, information systems and tool sets, and organizational constructs, and identify options needed to address the remaining gaps.

With these tasks in mind, the purpose of this monograph is to review current ACS planning, execution, monitoring, and control processes and identify where major gaps still exist between current practice and the 2002/2006 PAF enterprise command and control OA. We then outline implementation strategies to facilitate changes needed to achieve the TO-BE enterprise command and control OA.[7]

The Previous RAND Enterprise Command and Control Operational Architecture

Since the 1990s, the Air Force has supported nearly continuous deployments around the globe, engaging in varying operations, from small-scale peacekeeping and humanitarian relief operations to major combat operations (MCOs), such as Operation ENDURING FREEDOM (OEF) and Operation IRAQI FREEDOM (OIF). To support constant deployment, employment, and sustainment of Air Force forces, the enabling combat support system must be tailorable, flexible, and agile. Previous PAF analysis defined key elements of an ACS system (Tripp, Galway, Killingsworth, et al., 1999). The key elements include the following:

- an expeditionary, forward-thinking mind set, which would be instilled in combat support personnel
- recognition that ACS processes are essential elements of the Air Force command and control enterprise. These ACS processes would assess, organize, and direct combat support activities to meet operational requirements and respond to rapidly changing circumstances. The enhanced ACS capability would help combat support personnel do the following:
  - Estimate combat support resource requirements and process performance needed to achieve the desired operational effects for the specific scenario.
  - Determine feasibility of proposed operational and combat support plans. If combat support plans do not have adequate resources, develop mitigation strategies or initiate operational replanning activities.
  - Configure supply chains for deployment and sustainment, including the military and commercial transportation needed

---

[7] We discuss options but do not suggest a priority or develop costs associated with implementation plans.

to meet deployment and sustainment needs once a feasible plan is developed and adequate resources are committed to accomplish its objectives.

- Establish control parameters for the performance of various combat support processes required to meet specific operational needs.
- Track actual combat support performance against control parameters.
- Signal when a process is outside accepted control parameters so that plans can be developed to get the process back within control limits.

• a quickly configured and responsive distribution network to connect forward operating locations (FOLs), forward support locations (FSLs), and continental U.S. (CONUS) support locations

• a network of FOLs resourced to support varying deployment and employment timelines.

Command and control is the "brain function" of the combat support system. For example, meeting rapid-deployment operational requirements requires quick assessments of beddown plans so plans can be generated. Potential airfields' status and capabilities must also be analyzed. And the status of in-theater resources must be continuously updated to facilitate rapid time phased force and deployment data (TPFDD) development. Quick employment and subsequent sustainment require that theater and global maintenance and distribution operations be configured rapidly. Effective allocation of scarce resources requires the system to monitor resources in all theaters and prioritize and allocate them in accordance with global readiness. Finally, the system needs to be self-monitoring during execution and able to adjust to changes in either combat support performance or operational objectives.

Recognizing the importance of ACS, in 2002, the Deputy Chief of Staff for Installations and Logistics (the AF/IL at that time) asked PAF to develop an enterprise command and control OA. The resulting analysis described the Air Force enterprise command and control OA as it was in the 2000 time frame (AS-IS) and how it could be in the future (TO-BE) and outlined a framework for evaluating options for enhanced ACS. The analysis also described the changes that would need to take place to achieve that future state (Leftwich et al., 2002). With operations being conducted in Operation NOBLE ANVIL (ONA), followed by OEF and OIF, PAF further evaluated the principles of ACS, resulting in a refined and expanded enterprise command and control OA in 2006 (Tripp, Lynch, Drew, and Chan, 2004; Lynch, Drew, Tripp, and Roll, 2005; Mills et al., 2006).

Since that time, the Air Force has taken some steps to enable the implementation of the TO-BE enterprise command and control OA—for example, the establishment of the Air Force Global Logistics Support Center (AFGLSC) and establishing a war reserve materiel (WRM) global manager.

In addition, other influences, such as Program Budget Decision 720, dated November 7, 2006, reducing Air Force manpower end strength and budget cuts, changed the environment in which the Air Force operates. With end-strength reductions came increased centralized management of scarce resources. However, even as the defense budget tightens, the Air Force is being asked to support a full range of dynamic and irregular operations with limited resources. Integrating combat support capabilities through enhanced ACS is essential to the success of U.S. military operations.

Organization of This Monograph

In the chapters that follow, we present the results of our review of the TO-BE enterprise command and control OA and progress made by the Air Force in implementing it. In Chapter Two, we present the research approach and analytic framework, including how the operational environment has evolved since the 2002 enterprise command and control OA was developed. Chapter Three presents ACS planning, execution, monitoring, and control process shortfalls. Chapter Four codifies the ACS vision for the future, based on some core command and control principles. Chapters Five and Six describe how the analytic framework is or is not being applied in current ACS doctrine, training, tools and systems, and organizations. Chapter Seven concludes with some recommendations for improved ACS planning, execution, monitoring, and control.

In addition, the document contains the following three appendixes:

A. the RAND strategies-to-tasks framework
B. ACS annotated bibliography
C. joint and Air Force command structure.

CHAPTER TWO

Approach and Analytic Framework

This chapter begins by presenting the research approach used in this analysis (see Figure 2.1). It then outlines how the operational environment has evolved in the past several years.

Figure 2.1
Research Approach

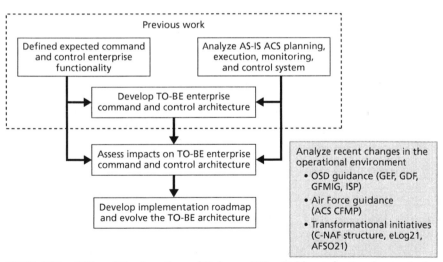

NOTE: OSD = Office of the Secretary of Defense. GEF = Guidance for Employment of the Force. GDF = Guidance for Development of the Force. GFMIG = Global Force Management Implementation Guidance. ISP = Integrated Security Posture. CFMP = Core Function Master Plan.

RAND *MG1070-2.1*

Research Approach

For this analysis, we begin by reviewing the PAF enterprise command and control OA developed in 2002 and expanded in 2006. We then examine the extent to which recent changes have affected both the AS-IS and TO-BE ACS planning, execution, monitoring, and control processes. For each area—process, doctrine, training, tools and systems, and organizations—we summarize the recommendations of the previous analyses and evaluate Air Force progress in addressing the issues. We then evaluate how changes in the operational environment, including changes in OSD planning guidance and Air Force transformational initiatives, affect ACS to determine the applicability of the 2002/2006 OA to today's military environment (changes are discussed individually in the next section). Next, we discuss the remaining shortfalls, and, finally, we present an implementation roadmap to help move the Air Force toward the updated TO-BE enterprise command and control OA.

The OA developed in 2002 identified issues and potential solutions that were, at that time, endorsed by Air Force senior leaders. Table 2.1 lists the key stakeholders with which we have worked since our initial analysis in 2002.

Over the years, incremental improvements have been made, but some issues still persist. Some of the TO-BE fixes identified in the 2002 OA have not yet been fully institutionalized. And the operational environment has changed. In the next section, we outline some of the factors of the current Air Force environment that could influence future ACS.

The Changing Operational Environment

The Air Force corporate structure today is not the same as it was in 2002. Changes in OSD guidance and new organizational structures have changed the environment in which the Air Force operates. These corporate changes could, in turn, change the TO-BE vision and the

Table 2.1
Key Stakeholders with Which We Have Worked During These Analyses

Air Force	Joint and Other Services
COMAFFORs: AFCENT, AFEUR, AFPAC, AFKOR, AFSOUTH, AFAFRICA, 18th Air Force	JCS, OSD
	USCENTCOM J-4, CDDOC
	USTRANSCOM/J-3/J-4/J-5, USTRANSCOM DDOC
AF/A4/7, AF/A4L, AF/A4I, AF/A4/7Z, AF/A4/7P, AF/A7C, AF/A7S, AF/A3O	USEUCOM J-4, EDDOC
SAF/XC	USPACOM J-4, PDDOC
MAJCOMs: AFMC, ACC, AMC, AFSPC, AFSOC	USFK J-4, PDDOC-K
AFGLSC	DLA
ALCs	Army
Operational wings	
AFIT	

NOTE: COMAFFOR = commander, Air Force forces. AFCENT = Air Forces Central.
AFEUR = Air Forces Europe. AFPAC = Air Forces Pacific. AFKOR = Air Forces Korea.
AFSOUTH = Air Forces Southern. AFAFRICA = Air Forces Africa. AF/A4L = Deputy
Chief of Staff for Logistics, Installations, and Mission Support, Directorate of
Logistics. AF/A4I = Deputy Chief of Staff for Logistics, Installations, and Mission
Support, Directorate of Transformation. AF/A4/7Z = Deputy Chief of Staff,
Logistics, Installations, and Mission Support, Directorate of Global Combat
Support. AF/A4/7P = Deputy Chief of Staff, Logistics, Installations, and Mission
Support, Directorate of Resource Integration. AF/A7C = Deputy Chief of Staff for
Logistics, Installations, and Mission Support, Directorate of Civil Engineering.
AF/A7S = Deputy Chief of Staff for Logistics, Installations, and Mission Support,
Directorate of Security Forces and Force Protection. AF/A3O = Deputy Chief of
Staff, Directorate of Operations. SAF/XC = Office of the Secretary of the Air
Force, Warfighting Integration and Chief Information Officer. MAJCOM = major
command. ACC = Air Combat Command. AMC = Air Mobility Command.
AFSPC = Air Force Space Command. AFSOC = Air Force Special Operations
Command. ALC = air logistics center. AFIT = Air Force Institute of Technology.
JCS = Joint Chiefs of Staff. USCENTCOM = U.S. Central Command. J-4 = Logistics
Directorate. CDDOC = USCENTCOM Deployment and Distribution Operations
Center. USTRANSCOM = U.S. Transportation Command. DDOC = deployment
and distribution operations center. USEUCOM = U.S. European Command.
EDDOC = USEUCOM DDOC. USPACOM = U.S. Pacific Command. PDDOC = USPACOM
DDOC. USFK = U.S. Forces Korea. PDDOC-K = USPACOM DDOC Korea. DLA = Defense
Logistics Agency.

way in which combat support processes are mapped in the enterprise
command and control OA.

Shifts in Office of the Secretary of Defense Guidance

Current planning guidance calls for increased operational requirements (DoD, 2007, 2008a, 2008b). DoD planning guidance for fiscal
year (FY) 2008 shifts the focus of the military from conventional mili-

tary operations toward irregular, catastrophic, and disruptive threats and capabilities while maintaining the ability to engage in two MCOs. Instead of focusing on two simultaneous MCOs, the focus shifts to maintaining homeland defense while also supporting ongoing steady state deployment commitments, which include a range of operations, from stability operations to irregular warfare and catastrophic attacks, all while maintaining capability to meet and defeat unforeseen challenges.

A parallel shift has also occurred in the roles Air Force forces play in these types of operations. More focus now falls on train, equip, advise, and assist operations in which Air Force personnel teach and aid host countries so those nations can become responsible for their own safety and security. In many of these cases, combat support capabilities rather than aircraft fighting missions might be the *tip of the spear.*

Still, the requirement to be ready to support two nearly simultaneous MCOs remains. To support MCOs in addition to these worldwide, nearly continuous steady-state operations, the Air Force might need to consider new ways to take better advantage of existing resources.

New Air Force Framework for Programming and Training

To support the warfighting mission as outlined in OSD guidance, the Air Force has a responsibility to organize, train, and equip forces. It must prepare its forces to support operational requirements as deemed necessary by the Secretary of Defense (SecDef), the JCS, and CCDRs. To better meet the training and readiness requirements, the Air Force has designated 12 service core functions as a way to present warfighting capabilities to CCDRs and link combat support functions to future programming needs.[1] ACS is one of the service core functions; an ACS CFMP is currently in development.

The ACS CFMP provides a guide for better integration of combat support activities—within the combat support community and with

---

[1]   The 12 Air Force service core functions are nuclear deterrence operations; air superiority; space superiority; cyberspace superiority; global precision attack; rapid global mobility; special operations; global integrated intelligence, surveillance, and reconnaissance (ISR); command and control; personnel recovery; building partnership; and ACS.

the operators. The goal is to coordinate all 26 core combat support functional capabilities to achieve specific operational objectives more efficiently and effectively. The operational objectives are to do the following:

- Ready the force.
- Prepare the battlespace.
- Position the force.
- Employ the force.
- Sustain the force.
- Recover the force.

The functional capabilities are as follows:

- acquisition
- airfield management
- air traffic control
- chaplain service
- civil engineer
- communications and information
- contracting
- distribution
- education and training
- financial management and comptroller
- health services
- historian
- judge advocate
- logistics planning
- maintenance
- manpower and personnel
- materiel management
- munitions
- Office of Special Investigations
- postal
- public affairs
- safety

- science and technology
- security forces
- services
- test and evaluation.[2]

This coordination is complex because combat support core functional capabilities are multiechelon and interrelated. The intent is for the Air Force to use the CFMP to "assess potential integration requirements and opportunities" (U.S. Air Force, 2010a, p. 1).

AFMC was designated the lead integrator for the ACS CFMP. Thus, AFMC now has ACS responsibilities. As part of our analysis, we consider how both the ACS and the command and control service core functions and their associated CFMPs could change the way in which ACS planning, execution, monitoring, and control processes are outlined in the TO-BE enterprise command and control OA.[3]

The Evolving Combat Support Enterprise

To respond to changing operational requirements, the Air Force is transforming its combat support enterprise—a very large system involving billions of dollars and tens of thousands of people. The transformational initiatives are large in scope and cover all aspects of combat support, including maintenance, distribution, procurement (sourcing), information, financial, and command and control activities. The Air Force has invested hundreds of hours of senior-leader time to set the direction, thousands of hours of staff time, and millions of dollars in specific transformation initiatives. The goal of the transformation is to enable ACS to meet emerging CCDR needs while also efficiently organizing, training, and equipping the force for day-to-day operations.

Such programs as eLog21 and AFSO21 aim to streamline and modernize ACS while also reducing operating and sustainment costs.

---

[2]   Both lists contain excerpts from U.S. Air Force, 2007.

[3]   Both ACS and command and control functions are influenced by ACS planning, execution, monitoring, and control, so both CFMPs could include ACS planning, execution, monitoring, and control processes. At the time this monograph was written, which service core function (ACS or command and control) would include ACS planning, execution, monitoring, and control processes was still under debate.

Centralization of the management of WRM (at ACC) and civil engineering unit type codes (UTCs) (at Air Force Civil Engineering Support Agency [AFCESA]) illustrates how the Air Force is moving toward a capability-based approach when making enterprise resource allocation decisions.[4] Other changes, such as USTRANSCOM being named the distribution process owner and the creation of DDOCs, also affect the combat support enterprise.[5]

With the scope of the transformational initiatives being so large, the individual objectives sometime overlap or even contradict one another and might not square completely with the future vision of the combat support enterprise. We consider these initiatives and the vision for the combat support enterprise as we evaluate the TO-BE command and control OA.

Integrated Agile Combat Support Management and Control Concepts

The Air Force is moving away from a commodity-centered focus in which each resource was managed separately by base or by theater.[6] With budget constraints and increased contingency-operations demands, not every location (base or theater) can maintain its own reserve of resources. Resources must be shared globally. The commodity-centered focus did not allow for an enterprise picture of global Air Force capability. The Air Force is now supporting the development of a capability-based approach to resource allocation decisions. This move better supports the uncertain global demands generated by the unpredictable nature of future demands for Air Force capabilities. This suggests the need for reevaluation of how combat support resources are managed and controlled in a resource-constrained environment so that decision-makers can assess how constrained resources could affect the ability to initiate and sustain contingency operations around the world.

---

[4] A UTC is a five-character, alphanumeric code that uniquely identifies a predefined standardized grouping of manpower or equipment to provide a specific wartime capability.

[5] The distribution process owner is responsible for coordinating and synchronizing the execution of the strategic distribution system.

[6] See, for example, several articles in *Combat Support C2* (2003).

Several PAF research efforts have shown the benefits of centralized management and control of selected materiel, including aircraft spare parts and end items and nonunit WRM (Peltz et al., 2000; Feinberg, Shulman, et al., 2001; McGarvey, Masters, et al., 2008; McGarvey, Tripp, et al., 2010). In part because of this body of research, the Air Force has centralized materiel-management responsibilities and decision authorities across aircraft spare parts into the AFGLSC as part of eLog21. There are not enough assets for every theater to have its own stockpile of resources. Resources must be shared globally to meet uncertain demands. To be able to effectively shift resources where they are needed, the Air Force has designated global managers for some resources. Munitions-management responsibilities have been centralized in the Global Ammunition Control Point (GACP), and nonunit WRM (including basic expeditionary airfield resources [BEAR]) is also in the process of being centralized—again, partly based on PAF research—under ACC. The Expeditionary Vehicle and Equipment Initiative includes establishing a virtual organization linking command equipment management offices, vehicle equipment management support offices, and MAJCOM vehicle management offices for centralized management (or assessment) of support equipment (and vehicles), and global managers have been designated for other end items, such as propulsion.[7] These concepts can play a key role in enhancing ACS processes and are considered as we evaluate the TO-BE enterprise command and control OA.

---

[7] The command equipment management office orchestrates the requirements determination process for equipment; the vehicle equipment management support office sets standards used in the vehicle requirements determination process.

Agile Combat Support Planning, Execution, Monitoring, and Control AS-IS Process Shortfalls

In this chapter, we begin by reviewing the process shortfalls that we identified in our 2009 gap analysis. Many of these gaps were identified previously in the 2002 and 2006 enterprise command and control OAs (Leftwich et al., 2002; Mills, 2006). We discuss Air Force progress in closing those gaps and consider the changing operational environment in which the OA would now function.

AS-IS Process Shortfalls and the Current Operational Environment

The shortfalls and gaps in the AS-IS or current ACS system fall into the following five major categories:

- poor integration of enterprise combat support inputs into operational planning
- inability to configure supply chain activities to achieve specific operational objectives, ascertain when performance falls short, and reconfigure the combat support infrastructure rapidly
- poor coordination of combat support activities with the joint services community
- absence of resource allocation arbitration across competing services and theaters
- inadequate understanding that combat support is broader than logistics.

When evaluating ACS processes in the current operational environment with the current organizational and doctrinal construct, we found that the same issues still persist. We discuss each shortfall in detail in the next several sections.

Poor Integration of Combat Support Inputs into Operational Planning

Similar to the problem operational planners face when trying to integrate air power with ground maneuver, operational and combat support planning often occur independently with little analysis of combat support feasibility in early course of action (COA) development. Currently, there is no standard process or format for operational planners to communicate key operational requirements to combat support planners (to feed beddown, TPFDD, munitions, spares, or transportation planning). Combat support planners projecting resource requirements and later planning the TPFDD often work from different assumptions and with information of varying degrees of fidelity regarding operational requirements. This can hinder timely, accurate combat support planning.

More often, combat support planners are given an operational plan and asked to generate the appropriate resources to support it without having the opportunity to influence the plan. Within the AOC, the Strategy and Combat Plans divisions seldom have resident combat support expertise. On the AFFOR staff, operational planners often develop COAs with little help from the logistics and installations directorates. Furthermore, many resources required to initiate and sustain combat operations are not within the COMAFFOR's area of responsibility (AOR). As a result, the AFFOR logistics staff does not have information on what global assets are available or what enterprise combat support processes can deliver. The small size of the AFFOR staffs also does not allow them to go to each global enterprise resource provider and determine the extent to which it has resources and processes needed to meet the specific operational plan objectives. As a result, the AFFOR staff ends up making assumptions about what the global combat support enterprise can deliver to the AOR. Thus, enterprise combat support capabilities and constraints are not considered in

COA development. This results in operational planners committing forces to actions with unknown risks concerning the sustainability of the operational plan.

As an illustration, an AFFOR staff combat support liaison officer might be on the COA development team, but he or she rarely has the opportunity to evaluate the plan's feasibility; he or she is simply asked to support the plan. Combat support feasibility is typically considered after an initial COA has been developed, not in parallel with development of the COA. This sequential process can result in combat support limitations being identified after a COA is complete from the operational standpoint. Then, the COA has to be reconstructed, delaying COA development, wasting time, and duplicating efforts.

For example, during ONA, operational planners chose the location for a potential beddown area without sufficient installations and mission support (A7) planning input. The resulting tent city had to be relocated because of flooding and encroachment on explosive safety areas. If combat support planning were better integrated with operational planning, the tent city–location issue might have been identified earlier during the planning process instead of after it was set up.

During OIF, combat support was closely integrated with operations, and constraints were factored into the planning process. As a result, substantial changes were made to the operational plan, and a supportable plan was developed before operations began.[1]

At the OSD level, DoD is updating the process used for contingency planning in Chairman of the Joint Chiefs of Staff Instruction (CJCSI) 3110.3C, *Logistics Supplement to the Joint Strategic Capabilities Plan*, by defining a logistics sustainability analysis (LSA) that integrates combat support feasibility into operational planning.[2] An LSA will be completed on a routine basis for plans when a TPFDD is generated to assess the logistical feasibility of the plan. Combat support constraints are identified during the planning process, so mitigation strategies can be evaluated as needed.

---

[1] Discussions with Deputy Combined Forces Air Component Commander, USCENTCOM, during OIF, June 2010.

[2] CJCSI 3110.3C is not available to the general public.

Likewise, the Air Force is revising its contingency planning processes to enhance the way in which the AFFOR staff conducts the LSA. In some areas—for example, fuels—the Air Force has business rules, tools, and systems that clearly support the LSA process. Integrated Consumable Item Support (ICIS), the fuels planning system, contains usage planning factors and consumption estimates.[3] For an LSA, AFFOR staff logistics personnel input the types and number of aircraft, their expected usage during the contingency, and expected beddown locations into ICIS, and the system calculates the fuels requirement by location. It is a well-defined and easy-to-use system employed throughout DoD to provide consistent requirements estimates. Other combat support resources, such as spare parts, services, and communications, do not have such well-defined, standardized, repeatable processes.[4] Often, experienced personnel using an ad hoc method provide their best estimates of requirements to support an LSA.

However, the LSA process is used only when a TPFDD is generated. There is no standard way to identify combat support resource requirements and quickly conduct combat support assessments during the adaptive planning process—the planning process used for operations like those identified in the ISP that are part of the ongoing steady state deployment commitments.[5] During adaptive planning, AFFOR staff personnel are often asked to estimate requirements based on their general knowledge and past experience. These personnel might provide good estimates; however, continuity of personnel is not guaranteed.

---

[3] ICIS is a DLA decision-support system that can calculate the deployment requirements for more than 2 million DLA-stocked items using a TPFDD.

[4] The AFGLSC has a process using the availability model within the Weapon System Management Information System to assess requirements for spares for any operations plan (OPLAN), but it has not been exercised frequently, and the techniques are understood by only a small complement of personnel. This process was used to evaluate one OPLAN and is currently being used to evaluate a second plan. The AFGLSC process can be incorporated into the LSA process as the requirements determination process for spares.

[5] *Adaptive planning* is the term now used for what was *crisis action planning*. The premise is that a crisis can occur quite quickly, from a variety of circumstances, and require any number of varied responses. Adaptive plans should be flexible enough to be applicable to multiple situations, to perceived or unknown threats, with slight modification.

The same personnel might not be in the same position five years from now. Or new personnel might have little or no experience on which to base an estimate. The requirements for the next adaptive plan could be generated in a different, ad hoc manner, possibly yielding a very different requirements estimate. The adaptive planning process is not well defined, making it difficult to integrate combat support and operations.

Further complicating integrated combat support and operational assessments are the many-to-many relationships the AFFOR staff has with individual resource providers. For example, the AFFOR staff might work with many different providers to secure needed resources, such as water in each country where forces might operate. Multiply this by the many necessary combat support resources, and that becomes a lot of providers with which a small staff must coordinate. Combat support is treated as a set of unrelated resources, making it difficult for the AFFOR staff to produce timely and integrated feasibility assessments for all resources. As a result, supply chain assessments for all commodities are not conducted either during COA development or during LSAs. Nor are individual commodity supply chain assessments—for example, munitions, fuels, spare parts, engines, and vehicles—balanced across commodities to identify the most-binding constraints and develop mitigation strategies across resource supply chains.

Not only do the stovepipes within the combat support arena affect operational planning, but the traditional separation between the combat support and operational planning communities might hinder effective integration. Most logisticians, for example, are not trained in and do not participate in air campaign planning. Combat support personnel have difficulty relating resource availabilities and process performance to needed operational capabilities within the planning cycle. They therefore have little understanding of how and when combat support considerations should play into the planning process. They are not skilled at communicating essential aspects and effects of combat support options in terms that are relevant to the operator.

For their part, operators lack logistics (including installation support) training and hence tend not to consider the likely effect that support capabilities will have on planned missions. When combat support information is not valued, combat support aspects of plans are more likely to be overlooked, resulting in overly optimistic operational plans that might have to be altered during execution because of combat support realities.

Inability to Configure Supply Chains to Achieve Specific Operational Objectives, Ascertain When Performance Falls Short, and Reconfigure the Combat Support Infrastructure Dynamically

To ensure that planned objectives are being met, each commodity supply chain needs to be configured to achieve the specific operational objectives outlined in a contingency or training plan. For instance, using the spare-part supply chain as an illustration, the transportation times, repair times, and supply levels at operating locations should be set to achieve specific sortie generation objectives as required in a COA or plan. This does not occur routinely today. Rather, predetermined levels are sent forward to operating locations with the units, called readiness spares packages (RSPs), but these levels are based on approved and funded planning factors, not on the specific factors being considered in the actual plan. In some cases, planning factors vary greatly from the OPLAN requirements, which can create a shortfall before operations even begin.

Furthermore, both combat support activities and operations should be continuously monitored for changes when actual performance differs from planned performance and appropriate resource providers, such as the AFGLSC or the GACP, alerted when actual performance can jeopardize planned operational missions. However, combat support feedback data, such as resource levels, rates of consumption, critical-component removal rates, and critical-process performance times (such as repair times, munitions buildup times, in-transit times, infrastructure capacity, and site-preparation times), might not be routinely recorded. Even when these data are available, they are typically the focus of planning and deployment rather than employment and sustainment. Because operations can change suddenly, these data should be continuously available throughout operations in order to make needed adjustments to the combat support system quickly and seamlessly.

To complicate the issue further, when system monitoring reveals a mismatch between desired and actual resource or process performance, it can be difficult to find the source. For example, for activities supporting multiple theaters (such as depot repair) or multiple services (such as a theater distribution system [TDS] or construction priority), the source of the discrepancies can be difficult to pinpoint. These discrepancies between desired and actual levels of support can arise from changes in combat support performance or from changes in operations. An assessment process should be used to address combat support performance problems quickly and estimate new combat support requirements to meet changing operational objectives. However, with limited monitoring and performance assessment currently being conducted, it is hard to know when to intervene and adjust combat support activities.

Poor Coordination of Combat Support Activities with the Joint Services Community

Most combat support activities entail coordination among the services and the joint services community. Examples include infrastructure repairs, fuels management, the distribution and storage of munitions and housekeeping sets, and transportation. Nowhere is such coordination more important and troublesome than in transportation and distribution management. Inter- and intratheater distribution rely on the combined efforts of the Air Force, Army, Navy, and commercial carriers, all of which have separate responsibilities and all of which depend on the others for successful operation. Nominally, the Air Force is responsible for providing airlift, the Army for providing ground transportation and port management, the Navy for providing sealift, and the CCDR for theater distribution, often through the appointment of one service component as the executive agent for managing distribution operations.

In principle, the distribution system can operate smoothly if all know their role and do their jobs; troubles can arise when the relative roles of the different contributors in an operation are not understood, expectations differ on anticipated performance, or priorities differ among the major players. Such was the case during OEF and OIF with the TDS. For example, the following issues arose:

- difficulty predicting cargo requirements
- difficulty configuring, reconfiguring, basing, and sizing TDS airlift
- uncertainty about appropriate metrics to judge airlift effectiveness and efficiency
- appearance of incomplete coordination among movement modes in meeting TDS needs
- incomplete visibility of cargo within the TDS
- artificial separation of the intertheater movement system from intratheater movements
- restriction of intertheater airlift assets from intratheater use in early phases of conflict
- inefficient use of intratheater airlift assets (Tripp, Lynch, Roll, et al., 2006).

In an attempt to improve those processes and deal with problems, the SecDef assigned deployment process ownership to U.S. Joint Forces Command (USJFCOM) and distribution process ownership to USTRANSCOM.[6] As part of executing its responsibilities, USTRANSCOM, with the consent of the commander of USCENTCOM, created a CDDOC in the AOR. The CDDOC works for the USCENTCOM J-4 and was created to improve the joint, multi-

---

[6] USJFCOM, as the primary joint force provider (JFP) for conventional forces, focuses on the global allocation of combat, combat support, and combat service support capabilities and forces to support combatant command requirements. Combatant commands, military departments, and the National Guard Bureau (NGB) provide force and capability commitment, availability, and readiness data to USJFCOM and its assigned service components. USJFCOM assesses the ability to sustain joint presence, operational commitments, and global surge capabilities over time based on allocation decisions and actions in effect. Recent DoD plans call for the closing of USJFCOM. When the command is dissolved, many processes, doctrine, roles, and responsibilities will have to be reassigned and rewritten. Although the command might be disestablished, the services will still have to integrate their individual capabilities to conduct joint operations.

modal, intratheater movement system and better integrate it with the joint, multimodal, intertheater movement system.

The Air Force relies on deploying quickly with small amounts of resources. This practice requires rapid resupply to sustain the forces. Because combat support depends on rapid and reliable transportation, TDSs should be structured to take full advantage of cooperation with the Army, Navy, joint services community, and coalition forces (if applicable). If rapid resupply cannot be established, the Air Force might have to rethink lean policies and deploy with more resources to sustain operations, which would lengthen deployment and employment timelines.

Just as combat support needs and capabilities should be communicated to operational planners, so too should they be communicated, agreed upon, and resourced with other services, the joint services community, and coalition organizations. In considering intratheater airlift, the Air Force should estimate transportation requirements based on anticipated sortie generation goals and understand the form in which those requirements should be communicated to the agency responsible for theater distribution. These estimates can be used to help structure demand-based distribution services.

Similarly, combat support personnel should clearly define capabilities to execute base beddown plans and be prepared to provide those requirements to coalition and allied forces that might host Air Force units in a contingency. Such communications with allied and coalition forces could accelerate site survey, base development, and beddown planning during the time-critical contingency planning process. They are essential to laying the foundation for coalition support and participation in execution of beddown and sustainment activities. They are also vital to how command and control of coalition installation support forces are established.

Absence of Resource Allocation Arbitration Across Competing Services and Theaters

Resources planned to support specific regions are sometimes diverted to support another AOR preparing for or engaged in an operation if so ordered within a service or through OSD. (A MAJCOM or C-NAF can allocate resources among units within a theater, but neither can formally allocate resources across competing AORs or between competing joint task force [JTF] demands across theaters.) When these cross-theater reallocations are made, there is little ability for rapid assessment of the effect that moving resources from one theater to another can have on readiness.

For example, the GACP at Hill Air Force Base (AFB) in Utah controls the global prepositioning and movement of munitions. It can, according to SecDef priorities, reallocate munitions from one AOR to another to support operations. Presently, however, if munitions are reallocated during an operation, no formalized assessment procedures exist to measure the readiness effect on both the giving and receiving AORs. Decision-support tools might exist, but a standardized process to reevaluate AOR readiness is not currently in place. Individual MAJCOMs and C-NAFs can assess munitions availability in their AORs using standard Air Force munitions computation models; however, showing impact across AORs is not as straightforward. During OEF, the 7th Air Force AFFOR staff plans and requirements directorate (the A5) and the A4 attempted to show how the reallocation of smart munitions from their AOR to operations in Afghanistan would affect other, existing contingency war plans in their theater. There was no defined process to help them show the effects and trade-offs during the contingency planning process.

OSD took some steps toward evaluating unit readiness by establishing the Defense Readiness Reporting System (DRRS) in 2002, which is a near–real-time readiness-reporting system designed to measure and report the ability of forces and support infrastructure to meet the requirements identified in the existing war plans and for the global war on terrorism.[7] It is capability-based using the Universal Joint Task List (UJTL) and Air Force Universal Task List (AFUTL) to identify the requirements for conducting contingency and other missions.

---

[7]   DRRS is the SecDef action in response to the National Defense Authorization Act for Fiscal Year 1999 added Section 117 to U.S. Code Title 10, which directed the SecDef to establish a "comprehensive readiness reporting system" that would "measure in an objective, accurate, and timely manner" the capability of the U.S. military to carry out the National Security Strategy, Defense Planning Guidance, and National Military Strategy.

However, DRRS does not have an analytic capability to show the effect of reallocating resources across war plans or across AORs. Although it provides a snapshot of readiness, it does not show how different resource allocation mixes would affect readiness. This type of readiness assessment should be completed as part of the contingency planning process before resources are reallocated so that high-level decisionmakers (up to and including the JCS) can see the effects of their allocation decisions before assets are moved.

These types of assessments should also be applied to the adaptive planning process. New OSD planning guidance predicts an increase in operational requirements. As the focus shifts from MCOs to maintaining homeland defense while also engaging in ongoing, steady state deployment commitments, more scarce-resource allocation decisions might be faced. Currently, there is no description of how combat support resources will be allocated and balanced from a global perspective.

Inadequate Understanding That Combat Support Is Broader Than Logistics

Attempts to incorporate combat support inputs into operational planning face not only the traditional separation between operations and logistics but also the separation between logistics and installation support. Logisticians and their installation support counterparts gain their experience and training along two very different career paths, and personnel are not well versed in each other's diverse activities.

Exacerbating the problem is the separation of the A4 and A7 functions in two directorates on the AFFOR staff. During OEF, rapid growth in base buildup and relocation motivated AFCENT to create an A7 installations support function. While providing AFCENT with senior-level, experienced decisionmakers in both logistics (the A4) and installations and mission support (the A7), it divided combat support between two organizations.

Analysis of ACS processes during ONA, OEF, and OIF showed duplication of some activities when these combat support functions acted independently but synergistic improvement when they teamed up. For example, initial AFCENT preliminary site surveys in logistics plans did not match up with engineer runway, parking, and infrastructure estimates. But, when the A4 logistics and A7 installation information was combined, the beddown planning proceeded smoothly. Another example was an A4 logistics dilemma with fuels off-load, flow, and storage at a few basing locations. When AFCENT logisticians integrated technically feasible COAs for solving the urgent fuels dilemma with inputs from ACC logistician and installations and mission support planners and from AFCENT installations staff, a mission solution was quickly identified and executed. Both examples illustrate the synergy of integrating A4 and A7 expertise during the planning processes.

AS-IS Process Shortfall Summary

Although progress has been made in improving ACS processes since 2002, there are still improvement actions that need to be taken. LSA processes and readiness data systems are being updated; however, at the C-NAF, operational planners do not always fully consider combat support feasibility when they develop plans. Many of the intended ACS planning, execution, monitoring, and control process fixes have not yet been fully institutionalized. Global managers are being established to manage and control scarce resources. But the process by which to allocate resources across competing demands has not been defined and written into doctrine. Each of the five shortfalls outlined in this chapter underscores the need for standardized, integrated ACS processes focused on operationally relevant results.

CHAPTER FOUR

The Vision for Meeting Agile Combat Support Planning, Execution, Monitoring, and Control Shortfalls

This chapter describes a vision, vetted with Air Force combat support leadership, for meeting the shortfalls discussed in Chapter Three. We first summarize the major elements of that vision. Then we present the vision in more detail as we discuss its theoretical underpinnings.

Key Elements of the Enhanced Agile Combat Support Vision

The ACS vision for addressing the shortfalls has three central elements. The first involves *creating standardized, repeatable processes to accomplish planning, execution, monitoring, and control of combat support activities within the Air Force command and control system to proactively manage scarce ACS resources across competing operational demands.* These processes are conducted by means of partnerships between the AFFOR combat support staff, global supply chain managers, global ACS functional managers, a global integration center (GIC), and the Air Staff.

The intent is to have the AFFOR staff concentrate on developing realistic demands for combat support capabilities (working closely with their CCDRs) while relying on the global combat support enterprise managers to identify enterprise capabilities and constraints, the second central element. Supply chain and ACS functional managers need *a way to integrate all the individual capability assessments and provide to the COMAFFOR an integrated set of capabilities that can be used in his or her contingency planning and execution actions, such as COA development.*

Finally, resource constraints are inevitable because of funding constraints being imposed on the program objective memorandum (POM) process. These constraints lead to the third component of the vision: *Processes for determining which CCDRs will have priority need to be developed between the SecDef, the JCS, and the Air Staff.* To provide leaders with the information they need to make tough trade-off decisions, standardized processes for identifying global resource shortages and operational outcomes associated with allocation of scarce resources should be established and defined in doctrine.

The Theoretical Underpinnings of Enhanced Agile Combat Support TO-BE Processes

Next, we discuss the theory that underpins the concepts that form the basis of ACS processes. These key concepts include the following:

- recognizing resource constraints in developing operational plans and allocating resources from a global perspective—using non-market economic principles and assigning supply, demand, and integrator roles and responsibilities
- assessing operational plans in terms of operationally relevant metrics—applying the strategies-to-tasks framework
- establishing feedback loops and control mechanisms—using cybernetics and electrical engineering control theory
- acting within the operational planning cycle—integrating combat support and operations processes.

The following sections define the key concepts and then illustrate how they provide operational-level ACS to CCDRs.

Recognizing Resource Constraints in Developing Operational Plans and Allocating Resources from a Global Perspective

Key ACS processes rest on principles discussed in the economics literature dealing with economies in which no market exists. This literature applies to situations in which there is not a market to help make resource allocation decisions. In a market economy, prices are used to align supply and demand for resources. In the world of operational planning, there is no such market mechanism to help in allocating scarce resources among competing consumers of resources.

To balance competing requirements (between the demand for combat support and the available combat support resources), we introduce some key concepts. First, we view the world in terms of *demanders* for resources, *suppliers* of resources, and an *integrator* who decides which demands will be satisfied when there are more demands than supplies (see Figure 4.1).

In using this economic perspective to view operational planning, we apply two principles to discuss these roles. First, supply-side, demand-side, and integrator decisionmaking processes should be independent of one another. If the integrator is too close to the supply side, decisions might be affected more by efficiency than by effectiveness. Operational needs might get inadequate attention. If, on the other hand, the integrator is too close to the demand side, current operations might always be given first priority, and efficiency might be ignored. We call this the *independence principle*.

Cases might occur in which the demand side supersedes the supply side. For example, operational requirements might supersede

Figure 4.1
Enhanced Agile Combat Support Processes Recognize Nonmarket Economics and Demand-Side, Supply-Side, and Integrator Roles

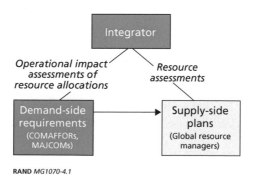

scheduled maintenance activities. If that is the case, then the supply side should know about future requirements and then develop flexible processes and take sustainment actions to accommodate requirements. In other words, if the independence principle is violated, those in senior leadership positions need to be aware that the principle is being violated and plan how to mitigate the potential resulting effects.

The second principle suggests that supply- and demand-side decisions should be made separately. According to this principle, the demand side specifies operational requirements and priorities and the supply side decides how to satisfy those needs—the demand side does not tell the supply side when and how to meet the operational requirements but rather when specific capabilities are needed. The supply side determines how to satisfy the operational requirements when needed.

When applying the principles of separation and independence in this economic framework, a tension results between the supply and demand sides. This tension is natural, and senior leaders need to recognize it. Once it is recognized, processes and organizations can be established to arbitrate between competing supply and demand requirements.

It is the integrator's job to arbitrate between the demand and supply sides. To arbitrate effectively, the integrator needs capability assessments to make informed trade-off decisions. Without these assessments, the integrator has limited visibility into the effects of his or her trade-off decisions. Capability assessments are key to making informed integrator decisions.

In addition, an Air Force integrator, or whomever is given the responsibility and authority to arbitrate between supply and demand, must function within the joint world and make recommendations to the SecDef and the JCS on the use of scarce resources, as shown in Figure 4.2. Current OSD planning guidance highlights the global, unpredictable nature of demands for Air Force capabilities (see DoD, 2007, 2008a). These uncertain demands might suggest the need for reevaluation of how combat support resources are managed and controlled in the face of limited resources, so decisionmakers can assess

Figure 4.2
The Secretary of Defense and Joint Chiefs of Staff
Set Priorities Among Combatant Commanders

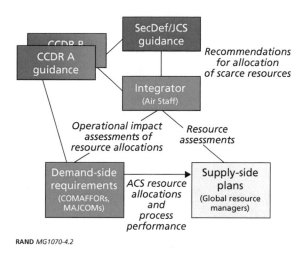

RAND MG1070-4.2

how constrained resources might affect the ability to initiate and sustain contingency operations.

For the foreseeable future, the Air Force, as well as other service components, will continue to have limited resources. Thus, we use this expanded strategies-to-tasks framework, which includes resource constraints, to identify shortfalls and suggest, describe, and evaluate options for implementing improvements in current ACS processes, doctrine, training, tools and systems, and organizations.

Assessing Operational Plans in Terms of Operationally Relevant Metrics

By making choices among CCDRs for scarce resources, the SecDef and the JCS are essentially indicating which demands will be satisfied. These choices will have a bearing on which operational effects might not be achievable in the various AORs, as shown in Figure 4.3. With the help of models or decision-support tools to inform the choices, decisionmakers could better assess and manage risk across competing demands. More work is needed to be able to relate combat support resource levels and process performance to operational effects.

Figure 4.3
Because Choices Are Made About Which Demands to Satisfy, Joint Effects Are Enabled or Constrained

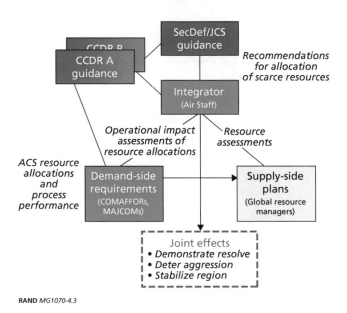

RAND *MG1070-4.3*

Operators have a long history of relating resources to operational effects. JP 3-0, *Joint Operations*, outlines how joint operational planning should tie military strategy to the employment of military power to achieve desired operational goals (JP 3-0, 2006 [2008], p. IV-1). Both the Joint Operation Planning and Execution System (JOPES) and the joint operation planning process (JOPP) require COA development as part of their planning processes.[1] One factor considered during COA development is effects—specifically, identifying desired and undesired operational effects.

A similar methodology—tying mission to objectives to tasks—is used in the development of the UJTL (CJCS Manual [CJCSM] 3500.04B, 1999b, Chapter 1). The JCS use the UJTL as a standard method for describing DoD capabilities. Based on mission

---

[1] JOPES is used prior to SecDef approval and Chairman of the Joint Chiefs of Staff (CJCS), transmittal of an execution order. The JOPP is used before and during execution of joint operations.

analysis, joint mission essential task lists (JMETLs) and agency mission essential task lists (AMETLs) were developed to help commanders identify tasks required for a mission to be successful. Both JMETLs and AMETLs have measures that focus on the outputs or results of performance of the task—the effect. Using the UJTL, commanders tie capabilities to operational effects.

By applying the UJTL and JMETLs and during the COA process, operators gain experience relating resources to effects. In both of these examples, military strategy and objectives are linked to operational tasks, similarly to how they were in the RAND strategies-to-tasks methodology developed during the late 1980s (see Kent, 1989, and Thaler, 1993). Again, more work is needed in this area. (See Appendix A for more detail on the RAND strategies-to-tasks methodology.)

While still improving the processes to integrate combat support constraints using the nonmarket, resource-constrained strategies-to-tasks framework, the analytic community does have many techniques that can relate combat support resource levels and process performances to metrics that are meaningful to operators. As shown in Figure 4.4, models are needed to relate combat support resource levels and process performances to metrics that operators understand, such as mission generation capability, FOL, initial operational capability (IOC), or full operational capability (FOC). Commanders of Air Force forces specify these metrics when developing their COAs.

To determine the combat support system's performance in terms of such capability metrics, it is necessary to understand how materiel and nonmateriel resources interact to produce the desired capabilities. Because these capability metrics depend on more than just materiel, materiel managers need to do more than simply monitor the numbers of physical assets available in each category; they also need to understand how asset location, condition, and quantities interact with repair, if applicable, and how transportation times in each category contribute to operational effectiveness. Ideally, those responsible for understanding combat support resources, including materiel and nonmateriel resources, would be able to relate the categories of resource—materiel, infrastructure, personnel, and transportation—to each other so the marginal contribution of individual resources can be determined

Figure 4.4
Combat Support Assessments Are Keyed to Stated Operational Goals

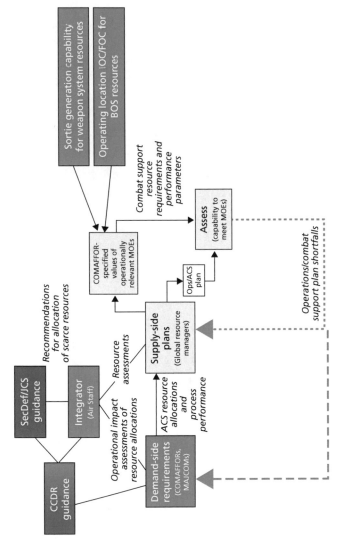

NOTE: MOE = measure of effectiveness. BOS = base operating support.

RAND *MG1070-4.4*

against systemwide operational-effectiveness output measures.[2] Decisionmakers would then be positioned to make the most cost-effective use of combat support resources: the use that maximizes capability to support the warfighter for a given set of resources.

Applying Feedback Loops and Control Mechanisms

The third key concept in the framework we used to evaluate ACS is based on cybernetic and electrical engineering control theory.[3] In electronics, sophisticated subsystems use three methods of control to create a stable system: proportional, integral, and differential control. The proportional control responds in proportion to how far off the system is from its control point. The integral control responds based on how long the system has been outside its control point. And finally, differential control responds based on how quickly the system went out of control. All three control methods depend on feedback to stabilize the process (Seborg, 1989).

Control theory is a concept that has been well understood in operational planning and has been the topic of operational planning doctrine for many years (Boyd, 1987). A closed-loop assessment with feedback and control mechanisms can inform operational planners of how the performance of a particular process affects operational capability.[4] For example, in operations planning, it is standard procedure to conduct battle-damage assessments and, if some targets have not been

---

[2] Air Force Instruction (AFI) 10-401 identifies requirements for conducting logistics sustainability analyses, including assessments of materiel, infrastructure (which is usually focused on FOL ramp, runway, and other construction needs), combat support forces (which is usually focused on personnel issues associated with filling combat support UTCs), and lift (which is usually focused on strategic and theater lift).

[3] In general, cybernetics is the study of the flow of information throughout a system and the way in which that information is used by the system as a means of controlling itself. In other words, a system based on cybernetic principles coordinates actions and decisionmaking while controlling the system but also senses changes in the environment and modifies itself to achieve its goals. See Beer (1966).

[4] A closed-loop process takes the output and uses it as an input for the next iteration of the process.

destroyed or rendered unusable, to modify the air tasking order (ATO) to retarget.

As shown in Figure 4.5, once a feasible plan is established, the jointly developed plan is then executed (the "Execute" box). In the execution portion of the process, actual performance of a combat support process is compared with the process-control parameters identified during the planning process, as shown in the lower right of the figure (the "Monitor/Control" box). When a combat support parameter falls outside the limits set in the planning process, combat support planners are notified so plans can be developed to bring the process back within control limits.

A key element of this closed-loop system is the feedback loop, shown by the output being fed back in as input, which determines how well the system is expected to perform (during planning) or is performing (during execution) and warns of potential system failure. This feedback loop, which includes feedback from senior leaders, tells the combat support planners when the plan should be reconfigured to meet dynamic operational requirements, both during planning and during execution.

Acting Within the Operational Planning Cycle

Enhanced ACS processes must be performed within the operational planning cycle, as shown in Figure 4.6. The figure shows the actions that must be performed in each phase of the operational planning cycle.

For example, MOE development in the ACS closed-loop system should be based on the priorities laid out in the CCDR guidance. If the CCDR's goal is bombs on target, then the ACS MOE might relate combat support resources to ability to generate sorties. Figure 4.7 shows a notional, fully armed sortie generation profile for a given weapon system that might be required to achieve desired operational effects over a campaign.

Figure 4.8 shows how three commodity supply chains might affect sortie generation capability in AOR X. The sortie-production capabilities in AOR X cannot meet the operational requirements as outlined in the contingency plan (CONPLAN), first because of ammunition and fuels limiting factors and then because of constraints on spares.

Figure 4.5
Once Feasible Plans Are Established, Supply Chain Performance Is Controlled to Achieve Operational Goals

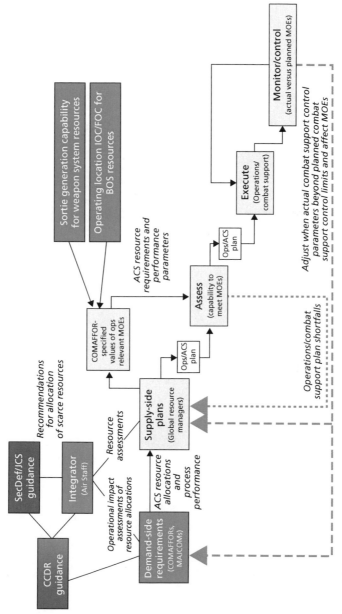

RAND MG1070-4.5

Figure 4.6
Enhanced Agile Combat Support Processes Need to Take Place
Within the Overall Operational Command and Control Planning
Cycle

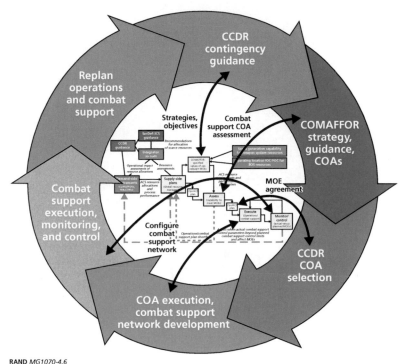

RAND *MG1070-4.6*

To meet the CCDR-specified sortie generation requirements, combat support capabilities can be reallocated from another AOR by a global resource manager. Figure 4.9 illustrates how sortie generation capabilities can be increased by reallocating spares from AOR Y to AOR X.

However, the sortie generation capability for AOR Y will decrease as a result of reallocating these assets. Figure 4.10 shows the original assessment of AOR Y's operational capability to meet wartime requirements. Figure 4.11 illustrates the effects on AOR Y's operational plan when spares are reallocated to AOR X.

The integrated closed-loop system, discussed above, could and should provide these assessments of proposed resource reallocations

Figure 4.7
**Combat Support Course of Action Assessments Focus
on Relevant Operations Metrics—for Example, Fueled
and Armed Sortie Requirements**

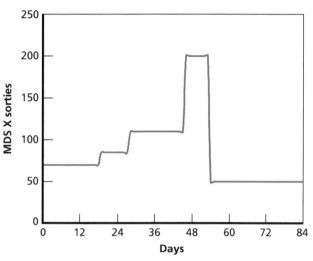

RAND *MG1070-4.7*

and how they affect other operational plans. Such assessments are necessary both in long-term planning and programming and in short-term execution. In the contingency planning process, such assessments can be used to quantify the effects of resourcing decisions for the Air Force corporate structure and the joint leadership. Effects of combat support funding can then be expressed in terms that the operational community understands, as was demonstrated in the PAF analysis mentioned earlier.[5] These analyses would also justify the guaranteed levels of support during execution: At a given funding level and with an understanding of global priorities, the global combat support system can guarantee support to a given operational plan at a clearly stated level. The flow of forces in the operational plan can then be informed by the stated level of support. In the adaptive planning process, such assess-

---

[5] In Amouzegar, McGarvey, et al. (2006), the authors show how the BEAR posture and resulting total system cost are affected by changes to the required delivery dates for BEAR assets at FOLs.

Figure 4.8
**Feasibility Assessments Determine Combat Support Constraints by Asset
for Area of Responsibility X**

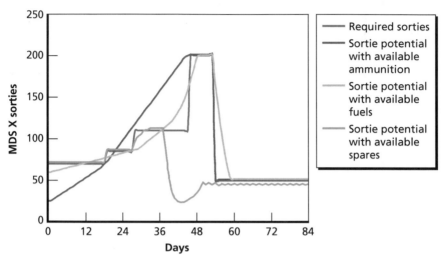

NOTE: MDS = mission design series.
RAND *MG1070-4.8*

ments can help the military leadership make tough decisions about how and where to accept risks from a global perspective.

However, with better assessments for allocation of scarce resources and enterprise management of assets comes potential risk. The systems used to perform capability analysis need to have security adequate to protect the information. The same information that is needed by Air Force planners could be used against the United States by enemy forces. Network vulnerabilities need to be evaluated. Redundancy and backups should be built into the command and control system, and methods should be developed for continuity of operations (COOP) when command and control is operating in a denied environment.

Figure 4.9
**Spares from Area of Responsibility Y Can Be Reallocated to Area of
Responsibility X to Meet Operational Requirements**

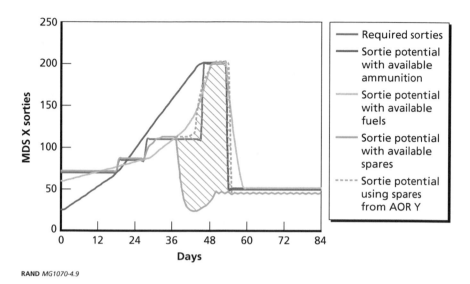

RAND *MG1070-4.9*

Balancing Supply Chains to Achieve Resourced Operational Objectives

As we indicated in Chapter One, this monograph covers combat support planning, execution, monitoring, and control. We now turn our attention to tracking and controlling supply chains to achieve operational objectives.

An ACS closed-loop feedback and control system needs to track actual combat support process performance against planned values. When the system breaches control parameter limits, the enhanced ACS system needs to signal combat support personnel that corrective action is needed. The 2002 TO-BE enterprise command and control OA outlines how this planning and control could occur across the echelons of support and throughout the phases of operational campaigns (Leftwich et al., 2002).

Currently, individual resources are managed and controlled independently, with little integration across categories of materiel. Man-

Figure 4.10
Area of Responsibility Y Operational Plan Assessment Before Reallocation of Spares Assets to Area of Responsibility X

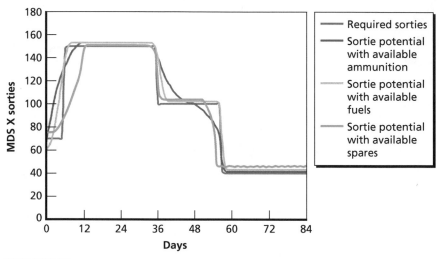

RAND *MG1070-4.10*

agement and control of these materiel resources need to be integrated with other categories of materiel, including WRM; vehicles and special purpose support equipment; munitions; petroleum, oils, and lubricants (POL); spare parts; and personal equipment, to determine how all materiel interrelates in terms of affecting operational objectives. Further, these resources need to be integrated with other combat support resources, such as civil engineering, communications, and security forces (see Figure 4.12). With constrained resources, leadership might have to make tough trade-offs. An integrated assessment across resources will provide an enterprise view of combat support on which leaders can base their decisions. In Chapter Six, we evaluate the organizational structure necessary to conduct these integrated combat support assessments for monitoring and controlling combat support resources.

Figure 4.11
Area of Responsibility Y Operational Plan Assessment After Reallocation of Spares Assets to Area of Responsibility X

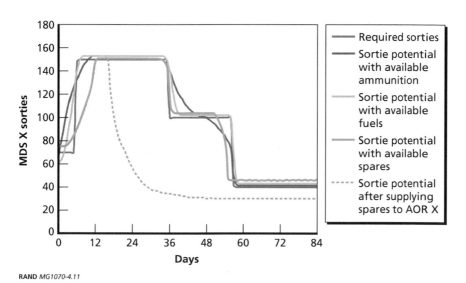

An Illustration of the Agile Combat Support Planning, Execution, Monitoring, and Control Processes

As discussed in the previous section, the combat support community does have the ability to relate many combat support resource levels and process performances to operationally relevant metrics (such as mission generation capability, FOL IOC, or FOC). For example, the sortie generation capability is a function of many combat support parameters, including removal rates of avionics components, maintenance throughput of the repair facility (both on base and at a repair facility), and movement capacity and throughput capability—for example, airlift frequency between the repair facility and a deployed location and transportation time for these components (see Figure 4.13). Degradation in any one of those combat support parameters will affect sortie generation capabilities, and the sorties projected might not meet the requirement.

Figure 4.12
Integration and Assessments Should Occur Within and Across Resource Categories

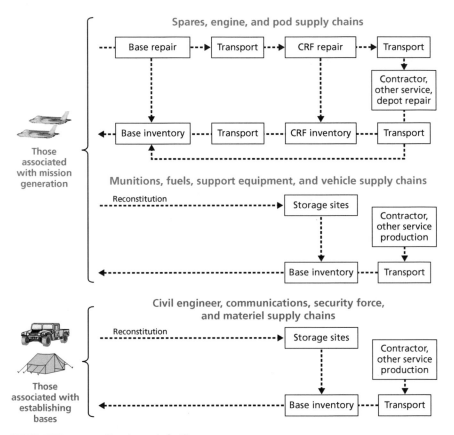

NOTE: CRF = centralized repair facility.
RAND *MG1070-4.12*

Combat support personnel should continue to assess and monitor the numerous metrics necessary to ensure that the support system can meet operational requirements, but the overall ability of the system to do its job should be reported in terms relevant to operators. For example, flying F-15s will result in avionics-component failure. These failed components will be replaced with serviceable components from the contingency centralized repair facility (CCRF) or another network repair facility (such as the depot), which requires the part to be trans-

Figure 4.13
How Movement Performance Is Related to Sortie Generation Capabilities

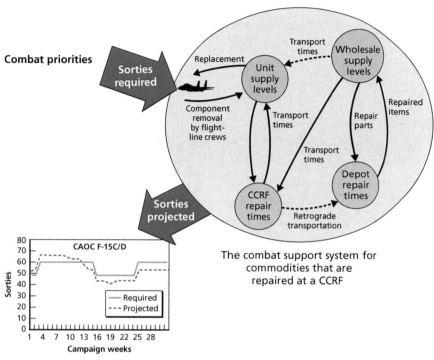

NOTE: CAOC = combined air and space operations center.
RAND *MG1070-4.13*

ported to the repair site. Components removed from the aircraft are also repaired at the CCRF or another network repair facility, and again the part must be transported to the repair site. The levels of avionics components deployed to the FOLs are determined by the transport time between the CCRF and the FOLs and the repair time for components at the CCRF. Each of these activities (and others) is shown at the bottom of Figure 4.14. Each should be tracked to ensure that the combat support repair network is correctly configured to support ongoing operations. However, these individual combat support metrics should be synthesized to report how they collectively can affect sortie generation—a metric of high interest to operators.

Figure 4.14
Combat Support Metrics Should Be Monitored and Assessed to Ensure System Performance

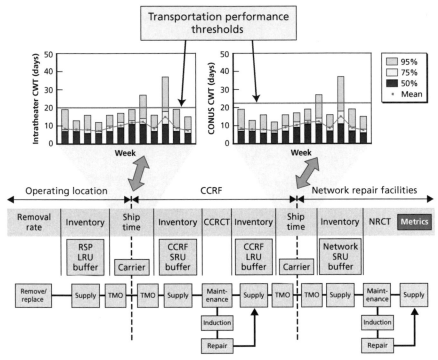

NOTES: CWT = customer wait time. CCRCT = contingency centralized repair center time. NRCT = network repair center time. LRU = line replaceable unit. SRU = shop replaceable unit. TMO = traffic management office.
RAND *MG1070-4.14*

A delay in setting up resupply pipelines could result in fewer sorties being generated, and yet there is no analytic process to translate added resupply time to a weapon system availability metric. As a result, combat support information tends to be provided to planners in the form of inventory levels or process performance (for example, resupply time) rather than base beddown capability, sortie generation capability, or other metrics more relevant to operational planning. Combat support personnel are not equipped to communicate combat support options in metrics that are meaningful to operators. And, for the most part, the tools to make this translation do not exist.

The combat support community can also develop a closed-loop ACS system with feedback to determine how well the system is expected to perform during planning and is performing during execution and then warn of potential system failure.

For example, the bottoms of Figures 4.14 and 4.15 lay out the closed-loop performance parameters for the combat support system shown in Figure 4.13. Data needed to track performance against each of these supply chain parameters are routinely collected, but they are not compared with performance levels needed to achieve specific operational objectives. This failure could stem from a lack of personnel who understand that the supply system contains wartime combat sup-

Figure 4.15
System Performance Should Be Compared with Expected Performance to Ensure That Operational Objectives Can Be Met

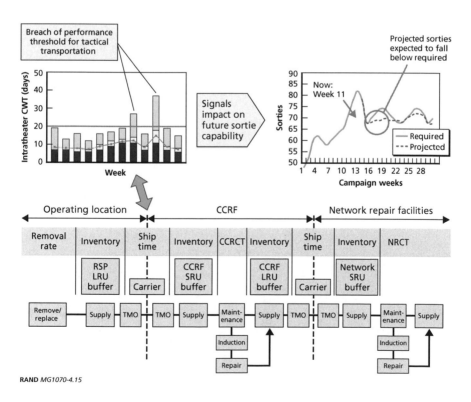

RAND *MG1070-4.15*

port performance parameters necessary to achieve specific operational objectives—for example, weapon system sortie generation objectives.

The top halves of Figures 4.14 and 4.15 show some data concerning transport performance consistent with data routinely collected by USTRANSCOM as part of its Strategic Distribution program. Such data can be obtained routinely from USTRANSCOM websites. The lines on both charts illustrate goals needed to support the availability goals of F-16s determined in the planning process.

Figure 4.15 shows that, when intratheater transport times breach the movement control limit (top left of the figure, where part of the bar extends above the horizontal line), it can be expected to affect F-16 sortie generation capability (top right of the figure, where the dotted line drops below the solid one at about week 15). Although technologies to make this type of assessment have been developed, this kind of analysis is not routinely performed.

Given the signal that intratheater transport times have dropped below the planned performance needed to support operational objectives, the AFFOR staff A4 can take several actions. The first is to approach the organization responsible for planning and executing TDS activities (the J-4) with these assessments and request support for improved transport service with the support of the combined force air component commander (CFACC). Requests could include suggestions for improvement—for example, increased frequency of service or alternative transport routings.

If transportation cannot be improved, then the AFFOR staff A4 and each deployed unit could take actions to readjust levels to compensate for the longer-than-planned distribution-system service. RSP levels could be changed by increasing the allocations from the ALC by changing the allocation among units within the theater or by changing the allocation of spares among units across theaters.

If transport service cannot be improved and there are not enough assets to adjust deployed-unit component levels, the operational plan might have to be adjusted. In this case, perhaps additional F-16s need to be deployed to the theater. Or some operational missions could be assigned to another type of aircraft. This is just one example of how a closed-loop ACS system could work.

Enhanced Agile Combat Support Process Summary

We used the key principles outlined in this chapter as a framework for identifying ACS planning, execution, monitoring, and control process shortfalls and suggesting potential mitigating options for the future. First, a nonmarket, resource-constrained strategies-to-tasks framework can help ACS planners better link combat support resources to their associated operational tasks by identifying supply, demand, and integrator roles and responsibilities. Metrics that are easily understood by operators should relate combat support resources and process performance to operational effects. Combat support personnel might need to continue to monitor each piece and pipeline within the system, but combat support parameters should be synthesized into metrics that are well understood by the operational community, such as sortie generation capability and FOL IOC, similar to the way in which operators track mission performance and report only deviations to the command post.

A closed-loop system, acting within the operational planning loop, with feedback mechanisms is essential to monitor and control combat support process performances during both the planning and execution processes. Combat support considerations should be integrated into initial COA development processes. A template to obtain operational data during adaptive planning, similar to the LSA process for contingency planning, is essential for proactive ACS engagement in the planning process. Controls should be implemented to signal when a system does not meet acceptable parameters so the combat support community can anticipate problems and so a plan can be implemented to get the system back inside control limits.

And finally, global management and control of combat support capabilities could facilitate resource allocation assessments across competing CCDRs, saving money and facilitating tough capability trade-off decisions. These assessments should be used to inform POM and other budgeting and program decisions. However, global management poses some risk of single-point failure. Geographically dispersed organizations with built-in redundancy can help, but methods to provide COOP and to minimize network vulnerabilities need to be developed.

In the next chapter, we apply the nonmarket, resource-constrained strategies-to-tasks framework to help guide the structuring of ACS doctrine, training, and information systems and tools.

Agile Combat Support Planning, Execution, Monitoring, and Control Doctrine, Training, and Information Systems and Tools: AS-IS Shortfalls and TO-BE Improvement Options

In this chapter, we address three areas—doctrine, training, and information systems and tools. As in Chapter Three, we begin by identifying shortfalls from the 2002 and 2006 enterprise command and control OAs and then discuss Air Force progression toward eliminating those shortfalls, as well as the changing operational environment, for each area in turn. We use the nonmarket, resource-constrained strategies-to-tasks framework discussed in Chapter Four to evaluate each area, suggesting changes based on the strategies-to-tasks principles.

AS-IS Doctrine Shortfalls and the Current Operational Environment

We begin with doctrine. The Leftwich et al. (2002) analysis revealed several ACS doctrine shortfalls. They include the following:

- ACS objectives and functions not well defined in doctrine
- lack of Air Force–wide emphasis on command and control for combat support
- combat support organizational responsibilities for command and control not well defined in doctrine or policy
- necessary command and control information flows not well documented for combat support.

Although doctrine clearly defines operational control (OPCON), tactical control (TACON), and administrative control (ADCON) of operational organizations, no such definition exists for combat support. Because doctrine is minimal for combat support, operational planning might not reflect combat support realities, delaying plan development, slowing the response to changing plans, and increasing the risk of running out of critical resources during extended operations.

Air Force doctrine and policy place little emphasis on combat support input to operational planning and execution. In AFDD 2, the joint services operations planning and tasking cycle phases are *plan*, *execute, assess*, and *adapt*. There is no feasibility check between planning and execution, so the process does not consider combat support status until after execution has begun. If plans are not supportable, corrective actions disrupt combat execution as well as future plans.

Because the ACS planning, execution, monitoring, and control concept is not well defined in doctrine, the objectives and functions of ACS and assignment of responsibilities to organizations are not well defined in policy. Supply, demand, and integrator roles and responsibilities should be defined so that, when trade-off or reallocation decisions for combat support resources are needed, those making the trade-offs are clearly charged with the responsibility and have the authority to make such decisions.

In contrast, operational command and control organizations are clearly defined. AFI 13-1AOC, Vol. 3, *Operational Procedures—Air and Space Operations Center*, provides guidance for the operation of the AOC and clearly denotes the functions involved in operational command and control.[1] It describes the purpose and primary responsibilities of the AOC, detailing the tasks necessary to accomplish them. It shows the command relationships between each division in the AOC, the information each division requires and generates, and the tools each uses to do its job.

Similar guidance for ACS planning, execution, monitoring, and control is largely contained in concepts of operations (CONOPS), which lack the directive authority of a doctrine or instruction document. MAJCOM and theaters develop operating instructions and

---

[1]   AFI 13-1AOC, Vol. 3, focuses primarily on regional air operations, not space, mobility, cyber, or global strike capabilities. These areas should also be enhanced in doctrine.

CONOPS independently, so the documents often differ from one command to the next in approach and process.

Although AFI 13-1AOC, Vol. 3, provides clear guidance for the operation of the AOC, it does not make explicit the information that operational planners should provide combat support planners outside the AOC (for example, combat support planners on the AFFOR staff) to drive timely and accurate combat support planning. This type of guidance is not provided in any Air Force policy or instruction.

AFDD 2-8 specifies four functions of a command and control system—planning, directing, coordination, and controlling—with little detail on the tasks necessary to accomplish these functions or which combat support organizations will perform them. Thus, there is confusion regarding the responsibilities of combat support organizations.

Potential TO-BE Doctrine Improvements

As part of this analysis, we reviewed more than 30 Air Force and joint documents, instructions, and policies relating to operational planning, command and control, combat support, and the future operational environment:

- GEF
- GDF
- GFMIG
- Annual Planning and Programming Guidance (APPG)
- ISP
- DoD Directive (DODD) 7730.65
- DRRS CONOPS
- JP 0-2
- JP 3-0
- JP 3-35
- JP 4-0
- JP 4-01.4
- JP 5-0

- JP 5-00.2
- CJCS Guide (CJCSG) 3501
- CJCSI 3110.01G
- CJCSI 3110.03C
- CJCSI 3141.01D
- CJCSM 3150.01
- CJCSM 3500.03B
- CJCSM 3500.04B
- CJCS Notice (CJCSN) 3500.01
- AFDD 1-1
- AFDD 2-4
- AFDD 2
- AFDD 2-8
- Air Force Policy Directive (AFPD) 10-2
- AFPD 10-4
- AFI 10-201
- AFI 10-401
- AFI 13-1AOC, Vol. 3
- AFI 10-244
- AFI 10-403
- ACS CONOPS.

Appendix B provides relevant annotations for each document. The purpose of the document review was to ensure that our TO-BE vision of enhanced ACS processes was supportable in light of current regulations; we found it to be so.[2]

In fact, many of the problems outlined in this section could be eliminated with a series of changes to Air Force doctrine and policy. In 2003, the Air Force initiated a review of its doctrine and policy, then started revisions to reflect the TO-BE enterprise command and control OA. Changes were implemented to AFDD 2-4, and it was planned that, as AFDDs 2, 2-6, and 2-8 came up for revision, they

---

[2] Recent DoD plans call for the closing of USJFCOM. When the command is dissolved, many of these documents will have to be updated, with the joint processes, roles, and responsibilities (those previously conducted by USJFCOM) assigned to other organizations.

would also include the ACS planning, execution, monitoring, and control concepts. Further, Air Force policy and procedures were to be written or modified in AFI (instruction) and tactics, techniques, and procedures (TTP) formats, to further detail the doctrinal concepts. However, many revisions remain incomplete. For example, AFDD 2-8 should include combat support details beyond the four basic functions of any command and control system. Currently, AFDD 2-8 does not address enhanced ACS concepts. The following are suggested inputs to AFDD 2-8 describing planning, directing, coordinating, and controlling.

Planning

Planning is defined as examining the environment, relating objectives to resources, and deciding on a COA. During both contingency and adaptive planning processes, it is critical to be able to add combat support information to initial planning processes, giving planners flexibility and confidence.

ACS functions include monitoring theater and global combat support resource levels and process performance, estimating resource needs for a dynamic and changing campaign, and assessing plan feasibility. Because capabilities and requirements change constantly, these activities should be performed continuously, so accurate data are available for COA and operational planning. Again, the data should be expressed in terms meaningful to the operational community—such as sortie generation and beddown capabilities.

Planning also includes assessment and ongoing monitoring of combat support infrastructure (FOLs; FSLs, such as WRM storage facilities; CONUS support locations, such as ammunition storage; the TDS; and command and control nodes) configurations that support the OPLAN. Benefits and limitations of various support options (sources of supply, transportation providers, modes and nodes, host nation support) should be weighed in the context of timelines, operational capability, support risk, and cost. Having complete, up-to-date information on FOL capacities and operational capabilities (for example, number and type of aircraft and munitions) and their support (for example, on-base repair capacity, fuels availability) allows more combat support information to influence OPLANs earlier in the planning process—that is, during COA development.

ACS processes should result in the production of a logistically feasible OPLAN, a combat support plan that dictates infrastructure configuration, a command and control organization structure, a TDS, and combat support resource and process control metrics.

Directing

Directing is defined as giving specific instructions and guidance to subordinate units. Directing activities for combat support include configuring and tailoring the combat support network and establishing process performance parameters and resource thresholds and buffers. Outputs from the planning process drive the direction of infrastructure configuration; there should be an ongoing awareness of combat support infrastructure and transportation capabilities to feed into operational planning and execution. For example, the speed and precision with which beddown sites can be assessed and prepared (configured) improve with the amount of information available beforehand. Awareness of the precise configuration for various options, in turn, gives planners more speed and flexibility in employment of forces in the face of changing objectives or constraints. The ability to quickly reconfigure the support infrastructure enables operational changes necessary as a result of anticipated or unanticipated changes in a scenario. Timely, accurate information and a combat support system able to execute network-configuration decisions would thus allow leaders to respond more quickly or simply to make more-informed decisions.

Along similar lines, identifying and using appropriate sources (for example, ships, supply depots, or host nation contractors) for different commodities (for example, ammunition, fuels, or spares) and required services (construction, billeting, feeding) allow maximum employment of available Air Force and joint services resources and the opportunity to balance intra- and intertheater requirements to support all AORs. As operational objectives change, requiring different logistics or installation support, the source can be changed. Also, as operational locations change, the source, as part of the overall combat support network, can change to meet the demands more quickly.

Coordinating

Coordinating is defined as sharing information to gain consensus, explain tasks, and optimize operations. Coordination in combat support ensures a common operational picture for all combat support personnel. It includes such things as beddown site status, weapon system availability, and sortie production capabilities. Coordinating should include monitoring ongoing operations and signaling when performance deviates from the given plan. ACS coordination activities should be geared to providing information to higher headquarters, not necessarily to seek a decision but to create an advance awareness of issues should a higher-headquarters decision eventually be needed. Combat support coordination tasks will affect theater distribution, force closure, supply deployment, and allocation of support forces. Each activity requires information gathered from a variety of processes and organizations and consolidated into a single decisionmaking framework that delivers accurate and complete data to planners and commanders.

For example, to coordinate TDS movements, combat support personnel should monitor all parts of the theater, as well as the activities of USTRANSCOM, other U.S. military services, coalition partners, and host nations. Similarly, base-level planning usually depends on supplies provided by intratheater distribution. To develop supportable plans, operational and support planners should understand what the TDS will provide at any given time. Policy should specify the information to be collected and dictate how it should be gathered and disseminated to organizations for decisionmaking or to maintain situational awareness.

Controlling

Controlling is defined as a composite function that uses parts of the planning, directing, and coordinating processes to ensure efficient execution of operations. During day-to-day and contingency operations, ACS tracks combat support activities, resource inventories, and process performance worldwide, assessing root causes when performance deteriorates, deviates from what is expected, or otherwise falls out of control. Control modifies the combat support infrastructure to return combat support performance to the desired state. ACS should evaluate the feasibility of proposed modifications before they are implemented and then direct the appropriate organizations to implement the changes.

Although doctrine should define and establish ACS planning, execution, monitoring, and control functions and objectives as described in this section, it should also prescribe which organizations perform these functions. AFDD 2 gives the organizational structure of the AFFOR and AOC, and AFDD 2-4 briefly describes the roles and deliverables of combat support functions. Doctrine should further delineate the roles and responsibilities of directorates within the AFFOR, divisions of the AOC, and other ACS nodes (see Chapter Six for further discussion of ACS nodes). It should include the reporting hierarchy and the communications network between groups. Once the what and who are delineated in doctrine, the AFIs should detail how the function will be executed, by describing tasks performed by each organization, the information that each group should consider in its decisionmaking, and how frequently this information is updated.

AS-IS Training Shortfalls and the Current Operational Environment

We now focus on the training and force-development issues identified in the 2002 enterprise command and control OA. The analysis revealed several ACS planning, execution, monitoring, and control training shortfalls:

- Most operations and combat support training focused on wing-level, not operational-level, skills.
- There was little training for ACS personnel on operational command and control and for operations personnel on ACS.
- There was little training on communicating and operating with the joint services community.
- Combat support participation in exercises and war games did not accurately address the execution planning process.
- There were few training opportunities.

The absence of well-defined supply, demand, and integrator processes, delineated in policy, contributes to a shortfall in training. For example, ineffective communications between operations and ACS planners can be attributed to the fact that combat support personnel typically do not have experience and are not taught their role in operational planning. As a result, they do not develop metrics appropriate for communicating with operators or the joint community.

Similarly, operators lack an understanding of how combat support contributes to and enables operational capabilities. They often set strategy without sufficient combat support input, which can lead to unsupportable or infeasible plans. War games and exercises do not focus on combat support requirements and often lack combat support realism. Operational planners generally do not consider combat support issues until well into the exercise-planning process, if at all, which can carry over into real-world practice, as it did during OEF.[3] If operators had better understood combat support requirements or if combat support personnel had been better able to communicate combat support capabilities, these issues might not have arisen.

This lack of awareness of each other's roles and processes and inability to communicate between operations and combat support become particularly evident in COA development. Combat support personnel describe their capabilities in terms of the amounts of fuels, munitions, and spare parts. Operations planners are more interested in assessments of combat support infrastructure that relate resources to FOL operating capability and sortie production. With proper training and enhanced education, this information could be incorporated into strategy at a much earlier point, but combat support planners neither know how nor have the tools to provide it.

We found that many numbered Air Force (NAF) staffs are inadequately trained in their management roles. Most NAF staffs are assigned to theater combat support roles from the wing level and have little or

---

[3]   In OEF, operational forces arrived well before their combat support and found their ability to fly missions hampered and their living conditions severe. And, although these lessons learned from OEF were applied during OIF, they have yet to be captured in doctrine. From discussions with Deputy Combined Forces Air Component Commander, USCENTCOM, during OIF, June 2010.

no experience in the diversity of combat support resource management at a theater level. In fact, of the current senior logistics leadership, only a very few members have demand-side experience and understand a systems view of combat support. In addition, assignment rotation does not allow personnel to become experts. Inexperienced personnel who are unclear about their responsibilities rotate in and ask the same questions as their predecessors did.

Many ACS personnel do not understand how to apply the non-market, resource-constrained strategies-to-tasks and closed-loop frameworks to maximize efficiency and effectiveness. More training and expanded educational opportunities are needed on relating combat support options to the CCDR's campaign plan to achieve joint operational effects. Little formal training is available to develop such skills. In fact, few opportunities for command and control training exist, leaving both operations and combat support personnel to learn their responsibilities on the job. Training on the job is problematic because manning for many command-level support functions at the NAFs is limited. Some on-the-job training is necessary, but, without supplemental information, it can reinforce bad practices and bypass issues that are not raised on a day-to-day basis. Examples of skills that are not formally trained include strategies-to-tasks, operational and combat support planning and assessments, managing the regional supply chain, nonunit sustainment and resupply resources, and theater-owned resources, as well as administering interactions between bases, MAJCOMs, headquarters, joint services forces, and the operations community. That, coupled with the absence of detailed policy, leaves many warfighting staff members and augmenters with little help in understanding of how to execute their responsibilities.

Potential TO-BE Training Improvements

Training shortfalls can be remedied through education. Training can be improved through the development of an ACS planning, execution, monitoring, and control curriculum, which can be incorporated into existing and new training courses, such as the joint services introductory course for basic AOC processes. This course can be expanded to include elements of a strategies-to-tasks framework, closed-loop systems, and operational-level ACS planning, assessment, and execution. This training should be encouraged and funded for ACS personnel with the same priority as it is for operational personnel.

In the longer term, enhanced ACS curriculum should train on such topics as combat support doctrine, policy, and guidance; AFFOR staff and AOC combat support processes; ACS capability assessments to incorporate combat support metrics into both theater and global capability measures; and new decision-support tools as they are developed.

The classroom instruction on the strategies-to-tasks and closed-loop methods and tools that does exist needs enhancement. The AFIT Logistics program could be expanded to teach the role of ACS planning, execution, monitoring, and control within a systems view of military logistics. Current AOC training at Hurlburt Field could also be expanded. Taking advantage of the expertise in these training units, expanded training could include testing new tools, systems, and processes before they are fielded. Strategies-to-tasks education could be provided through Air Force continuing education and could emphasize planning in both contingency and adaptive environments.

Exercises and war games should include more combat support issues and be funded to educate both operators and combat support planners on their respective roles and the role of combat support resources in campaign planning. The Air Force should take advantage of joint services logistics war games to evaluate new concepts and expand ACS skill training in tactical-level exercises.

Career-path planning for combat support personnel might include assignment to warfighting command-level positions in supply, transportation, logistics plans, civil engineering, or services, with the intent of creating senior combat support personnel with the skills needed to fill AFFOR staff A4 and A7 and CCDR joint staff ACS positions. Those combat support officers with a strong command and control background can be groomed for leadership positions. Additional education and training might be needed for those who will occupy key ACS assignments that are responsible for integrating combat support into the joint system, such as in the CCDR J-4 staff, the COMAFFOR

A4/7 staff, and the AOC. The number of positions is not large, but the positions are key to the development of feasible operational plans.

It is important that some of the best ACS personnel—promotable colonels—be assigned to demand-side organizations. Once an ACS officer has demand-side experience, he or she can rotate back to a supply-side functional area or assume an integrator position at a different level.

Finally, the role of combat support planners during COA development for both adaptive and contingency planning needs to be defined in doctrine and policy, trained to and exercised during peacetime, and implemented during operations. The Air Force should ensure that operators are trained to create operational planning teams, in a timely manner (understanding their uncertain planning environment), that include combat support planners. Operators should understand what combat support planners need and when, and combat support planners should understand the limitations and uncertainties within which the operators work. Only by training both groups to understand both sides of the planning equation and to communicate effectively will this link between operational and combat support planning be forged and sustained.

An important piece of the combat support planning process is the feedback loop, which enables combat support input to affect operations planning. For the feedback loop to be most effective, combat support personnel should understand air campaign planning and aerospace force capabilities. For example, what issues factor into planning different phases of the air campaign? What factors drive weapon system and preferred-munitions selection? What other weapons can provide similar effects? The combat support planner of tomorrow, working side-by-side with operations planners in an integrated planning process, should be able to answer these questions. Changes to training and improvements in education should equip combat support personnel to translate combat support resources to operational capabilities.

Training and realistic exercises are critical aspects of the link between combat support and operational planning. Educating both combat support and operations personnel about their roles in the context of campaign planning will enable more-effective communications and facilitate the integrated decisionmaking process in the TO-BE architecture.

AS-IS Information System and Tool Shortfalls and the Current Operational Environment

Finally, we evaluate information systems and tools. The shortfalls identified during the 2002 analysis include the following:

- tools needed to
  - relate operational plans to combat support requirements
  - convert combat support resource levels to operational capabilities
  - aggregate capability assessments to a theater or global scale
  - conduct capability assessments and aggregate them on a theater or global scale
  - conduct trade-off analyses of operational, support, and strategy options
- inability to access data on a timely basis
- proliferation of tools and systems, which has resulted in marginal success in fielding capabilities.

As expected, current information and tool shortfalls are consistent with current process shortfalls. Existing tools do not provide an integrated view of combat support process performance. Demand-side tools cannot quickly relate combat support capability to force packages, while supply-side tools to help execute global supply chain decisions have not been implemented. Tools are needed to convert OPLANs and status to combat support resource requirements and resource levels and then into operational capabilities, and systems are needed to monitor combat support capacity, resource inventory, and process performance levels. And tools to monitor actual process performance against plans and to signal when the performance falls below a significant threshold also need to be developed.

Additionally, tools are needed to determine enterprise repair capacities and capabilities needed to support contingencies. Tools are also needed to inform maintenance workload decisions by expressing infrastructure status in terms of operational capabilities and estimating resupply, beddown, and associated sustainment requirements. These tools will enable the Air Force to express its resupply and sustainment needs more accurately. Finally, tools are needed to aid beddown decisions. Some of these requirements can be supported by integrating and modifying existing systems, whereas others will require new system development.

Existing ACS planning, execution, monitoring, and control systems cannot seamlessly assess, prioritize, and reconfigure combat support resources, primarily because of the lack of uniformity among systems. Because combat support resources have been managed and funded by commodity, with different organizations having commodity-management responsibility, corresponding information systems have been developed and implemented independently for the organizations. The result is a myriad of independent systems with little ability to share data or interface with other systems. Thus, although these systems allow individual commodity data to be recorded and monitored, they do not facilitate the integration of the data for comprehensive combat support resource monitoring and capability assessments.

Even where good combat support assessment tools exist, they are unique to that command and not readily accepted by or interoperable with other MAJCOMs. Thus, the combat support system has difficulty making the appropriate assessment quickly and adapting accordingly.

Existing information systems often contain dated information and are therefore unreliable. Reliable recording of time-sensitive and often classified data within a globally distributed mobile organization, such as the Air Force, is inherently challenging. For example, logistics planning factors, which govern the translation of operational plans to combat support resource requirements, are updated only every few years. Similarly, base and host nation infrastructure capacity is updated only as needed. These factors result in combat support plans that are not reliable.

Potential TO-BE Information System and Tool Improvements

The Air Force already had tools that performed some of the recommended functions described in the previous section. In the near term, these tools can be leveraged to provide enhanced information and data for combat support planners. The following are merely examples of these; certainly, more are in various stages of development and are too numerous to be described here.

On the demand side, having a TPFDD planning tool that could rapidly capture combat support demands to support specific operational requirements would be key to better integrating ACS within the operational OODA loop. For example, a system that could provide the means to calculate the manpower and equipment required to satisfy operational requirements, to source and time phase each, and then to assess the transportation feasibility of each plan, all in a very short time, could be employed early in the planning stages to help with preliminary planning requirements. Existing information systems are unable to support these capabilities rapidly, although pieces of such an information system do exist within and outside the Air Force. A suite of tools that could automate as much of this planning work as possible would greatly expedite the contingency and adaptive planning processes.

A 2004 PAF analysis produced a prototype of a requirements TPFDD generator (Snyder and Mills, 2004). The model, Strategic Tool for the Analysis of Required Transportation (START), generates UTC requirements as a function of rules—for example, number and type of aircraft beddown, beddown conditions, and threat conditions. Although a few years old and in need of rule and database updates, a fully developed version of this tool could enable the kind of quick planning processes, early in the planning cycle, envisioned in the TO-BE OA.

On the supply side, some prototype allocation tools are already available for ACS use. Mixed-integer programming tools and data have been developed to source WRM from global sites, and tools have been developed to determine when to activate CCRFs (McGarvey, Tripp, et al., 2010). Mathematical relationships and formulas have been developed to relate aircraft availability objectives and sortie generation capability and the support parameters. Deploying units use the Dyna-METRIC Microcomputer Analysis System to determine levels of avionics components to take on deployments to meet specific aircraft availability objectives, given the repair concept and expected resupply (transport and processing) times.[4] In essence, this wartime spares computation system contains the combat support war plan and planned combat support performance parameters needed to meet the operational availability objectives called for by the CFACC.

The Execution and Prioritization of Repair Support System Planning Module, an Air Force–developed system, and Advanced Planning and Scheduling, a commercial off-the-shelf system, are two closed-loop spares planning and control systems that link depot processes and constraints to AFMC's spares planning process. Both can predict customer needs, prioritize them, and evaluate depot resource availability. Further analysis should be completed to see whether these systems will meet future Air Force needs.

A commonly described adaptive planning shortcoming was that operators had to plan with incomplete combat support data. As a result, aspects of plans were often made based on outdated information and assumptions, with the combat support information typically requested piecemeal as it became necessary. Currently, Air Force capability assessments are done ad hoc for each contingency in each theater. Although some customizing for COMAFFORs is necessary, combat support planners often reinvent the wheel when it comes to capability assessments. The content and format of these should be designed rationally and codified. This will enable personnel to be trained consistently and to think and communicate in the same terms across nodes. The frequency with which these assessments are done should be analyzed and standardized, too.

Any tool enhancements undertaken by the Air Force need to provide an integrated view of combat support resource allocations and

---

[4]   For more on METRIC (predecessor of Dyna-METRIC), see Sherbrooke (1966). For more on Dyna-METRIC, see Hillestad (1982).

process performance. Currently, information about Air Force resource and process metrics is often organized by commodity or end item and located on disparate information systems. Creating a single information system, such as the planned Expeditionary Combat Support System, that is accessible to a wide audience would enhance the visibility that leaders have over these resources. Such an information system would need to have enough automation to translate lower-level process and resource data into aggregated metrics and even some operational metrics (for example, weapon system availability or sortie generation capabilities). This information system could inform commodity managers, planners, and senior leaders who must make decisions during operations in which changes occur rapidly.

In the near term, prototype tools can be used for enhanced ACS assessments. For the long term, a thorough evaluation should consider all decision-support tools for a particular function, with implementation focused on a smaller set of tools worldwide. This will reduce the number of systems and training programs required for each planning function and permit an efficient transfer of information. New tools should be built on a systems infrastructure that can rapidly transfer information to all key combat support nodes, as well as the AOC and all relevant operational nodes. This infrastructure will maximize the productivity of new tools and allow them to interface with joint services systems.

The effects of improved information systems and decision-support tools will be felt throughout the TO-BE process. Properly integrating information from these tools will greatly reduce the chances of needing to revise a plan in midstream, allow a faster transition to operations and better-informed decisions, and facilitate change when necessary.

Agile Combat Support Planning, Execution, Monitoring, and Control Doctrine, Training, and Information System and Tool Summary

Many of the problems outlined in this chapter could be eliminated with a series of changes to Air Force doctrine and policy. Elevating the importance of enhanced ACS in Air Force doctrine and delineating roles and responsibilities using the strategies-to-tasks framework would provide enforceable rules for each organization, document information to be shared, and enable an improved planning process.

The absence of well-defined supply, demand, and integrator processes, delineated in policy, contributes to a shortfall in training. Many ACS personnel do not understand how to apply the nonmarket, resource-constrained strategies-to-tasks and closed-loop frameworks to maximize efficiency and effectiveness. More training and enhanced education are needed on relating combat support options to the CCDR's campaign plan to achieve joint operational effects. With few opportunities for command and control training, most combat support personnel learn their responsibilities on the job.

Training could be improved through the development of an ACS planning, execution, monitoring, and control curriculum (which includes elements of strategies-to-tasks; closed-loop systems; operational-level ACS planning, assessment, and execution; ACS doctrine, policy, and guidance; AFFOR staff and AOC combat support processes; enhanced ACS capability assessments to incorporate combat support metrics into both theater and global capability measures; and new decision-support tools) and career-path planning for combat support personnel.

Current information and tool shortfalls are consistent with current process shortfalls. Demand-side tools cannot quickly relate combat support capability to force packages, while supply-side tools to help execute global supply chain decisions have not been implemented. In the near term, existing Air Force systems and prototype tools can be used to provide enhanced information and data for ACS planners. In the longer term, enhanced ACS tools and systems should maintain an integrated view of combat support resource allocations and process performance. Properly integrating information from these tools will greatly reduce the chances of needing to revise a plan in midstream, allow a faster transition to operations and better-informed decisions, and facilitate change when necessary.

Finally, the role of combat support planners during COA development for both adaptive and contingency planning needs to be defined in doctrine and policy, trained to and exercised during peacetime, and implemented during operations.

We now turn to the organizations that perform ACS planning, execution, monitoring, and control processes as envisioned in the TO-BE OA. In the next chapter, we use the nonmarket, resource-constrained strategies-to-tasks framework to guide organizational structure options.

Agile Combat Support Planning, Execution, Monitoring, and Control Organizational Structure: AS-IS Shortfalls and TO-BE Improvement Options

We begin this chapter by identifying shortfalls from the 2002 and 2006 analyses in ACS planning, execution, monitoring, and control organizational structure. We then discuss Air Force changes in organizational structure since the analyses and then address future options using the nonmarket, resource-constrained strategies-to-tasks framework as a guide.

AS-IS Organizational Shortfalls and the Current Operational Environment

The Leftwich et al. (2002) and Mills et al. (2006) analyses revealed several ACS planning, execution, monitoring, and control organizational shortfalls, including the following:

- lack of clarity in warfighting roles and responsibilities when transitioning from steady state operations to contingency operations
- minimal staffing of warfighting organizations that relies on poorly trained augmentees[1]
- the difficulty that peacetime organizations have shifting to support one AOR from another

---

[1] For more information about organizational issues in transitioning from steady state to contingency operations, see Appendix C.

- lack of clarity in roles of joint services and combatant commands (COCOMs)
- the fact that resources are managed by different organizations.

As pointed out in Chapter One, since the 2002 and 2006 analyses were published, the Air Force has made progress in addressing the first two issues listed above. The Air Force has taken action to consolidate and standardize AFFOR staff organizations to meet the Air Force warfighting responsibilities. These AFFOR staff organizations will continue to evolve as COCOM needs change and as fiscal pressures continue to create the need to continuously evaluate how the Air Force can best meet its warfighting responsibilities.

The Air Force has also worked to establish formal relationships with Air National Guard (ANG) and Air Force Reserve Command (AFRC) units to augment the AOC and AFFOR staffs. A lot of attention has been given to AOC augmentation; AFFOR staff augmentation needs are now also receiving needed attention. Augmentation units are being assigned to help meet operational requirements. In addition, the 505 Command and Control Wing has initiated an AFFOR staff training course.

Although there is much more to do in the these areas, especially within ACS planning, execution, monitoring, and control, the Air Force understands what is needed to better support AFFOR staffs and has programs and policies to address these needs. The last three bullets in the list above, on the other hand, still require significant attention. With this analysis, we investigate in more detail organizational shortfalls that include the lack of

- an enterprisewide perspective when organizations are managing their resources
- an enterprise organization that has visibility of ACS manpower and equipment resources and the analytic capability to identify global ACS resource constraints to COMAFFORs—currently nonsupportable COAs can be generated and presented to joint services and CCDRs

- a formal organization to seek priorities for allocating scarce resources among competing AORs to achieve supportable operational objectives.

As we pointed out in Chapter One, the Air Force has acted to address some of these shortfalls by creating ACS organizations with enterprisewide responsibilities. Examples include the creation of the AFGLSC to manage the complete supply chain for component spares, support equipment, and vehicles; the GACP has enterprisewide responsibilities for the management of enterprise munitions, tanks, racks, adapters, pylons, and related assets; and ACC Plans and Integration has global responsibilities for WRM and BEAR management. In addition, the Air Force has functional area managers (FAMs) who are responsible for developing ACS personnel skills and career-path advancement.

These efforts are a step in the right direction, and they have improved visibility of ACS manpower, equipment, and other materiel; however, they have stovepiped resource responsibilities. The Air Force lacks an enterprise organization with the analytic capability to identify global ACS resource constraints, including the ability to identify the most-binding constraints with respect to specific COAs. COMAFFORs, as a result, might be presenting COAs to joint services and COCOMs that are not supportable from a global-resource point of view. In addition, the Air Force has not formally designated an organization to seek priorities from the SecDef and the JCS for allocating scarce resources among competing AORs to achieve supportable operational objectives. And, as we noted in the previous chapter, doctrine and policy do not clearly define and delineate the command and control roles and responsibilities of combat support organizations. When viewed in terms of the strategies-to-tasks framework, supply, demand, and integrator roles are not clearly defined and assigned to separate ACS organizations.

ACS planning, execution, monitoring, and control processes remained fragmented for several reasons. For example, on the demand side, as previously discussed, AFFOR staffs have many-to-many relationships with individual supply-side resource providers within the service, between services, and among national-level providers (see

Figure 6.1). As the figure shows, each NAF has to coordinate with many WRM locations and sources of supply and repair. These numerous relationships make it difficult to state systemic resource requirements in a timely manner.

Similar issues persist on the supply side. Many processes are fragmented, with incomplete process assignment to standing or existing organizations. There are multiple supply chains, each of which can be stovepiped and separate. The fact that different commodities fall under the responsibility of different organizations complicates combat support resource assessment. Although commodities have different characteristics that can dictate that they be handled and managed in distinct ways, they still need to be viewed from the perspective of how, in concert, they affect weapon system combat capability. Data are recorded in separate information systems; policies and procedures vary for each organization; and decisions are made on an individual commodity basis rather than from a comprehensive support perspective. This *stovepiping* of decisions affecting resource prioritization can lead to an imbalance between desired and actual capability and could misrepresent available capabilities.

And, operational integrators (AF/A3O and Directorate of Operational Planning, Policy and Strategy, Deputy Chief of Staff for Operations, Plans and Requirements [AF/A5X]) are defined in AFI 10-401 for the sourcing solution process; however, no coequal combat support integrator has been established. Combat support inputs might be funneled through the FAM, but no formal integration partnership has been formed with the combat support community. As previously stated, these problems have a deleterious effect on overall system and operational efficiency.

Potential TO-BE Organizational Options

The ACS organizational construct is large and encompassing, with many different supported and supporting organizations (see right side of Figure 6.2). However, the ACS planning, execution, monitoring, and control construct does not work independently. It is an integral

Figure 6.1
Support Coordination Requiring Many-to-Many Air Force Forces Staff Relationships

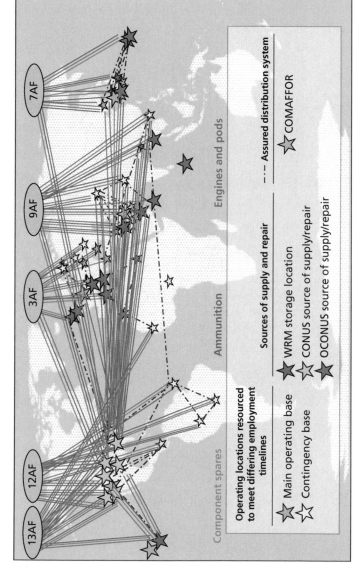

NOTE: OCONUS = outside the continental United States.

RAND MG1070-6.1

Figure 6.2
The Agile Combat Support Planning, Execution, Monitoring, and Control Structure Works with and in Support of the Air Force and Joint Command and Control Structure

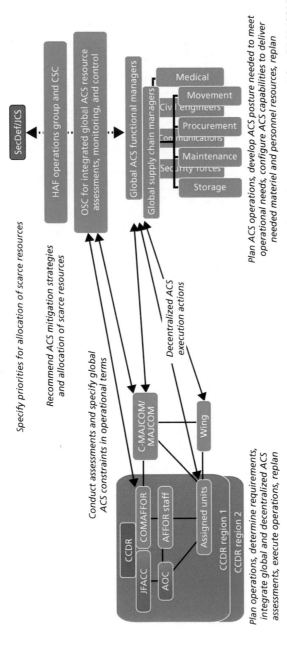

NOTE: HAF = Headquarters Air Force. OSC = operations support center. JFACC = joint force air component commander. C-MAJCOM = component MAJCOM. This figure depicts a TO-BE vision of Air Force command and control. Joint organizations are shown in purple. Air Force organizations are shown in blue. This figure is similar to the chart used by the Air Force to describe ACS planning, execution, monitoring, and control processes in Wickman and Battles (2009, slide 3). The difference is that this figure shows the future vision (TO-BE) for Air Force command and control, including some processes that are currently not assigned to organizations. These processes are discussed in detail in this chapter.

RAND *MG1070-6.2*

part of the Air Force enterprise and joint command and control structures (shown on the left side of Figure 6.2) and should work in conjunction with Air Force and joint command and control.

An important step toward resolving the ACS problems identified in current and past analyses is to establish a standing ACS organizational structure with ACS planning, execution, monitoring, and control processes assigned to organizations with clearly defined roles and responsibilities. Applying the nonmarket, resource-constrained strategies-to-tasks framework within the current Air Force enterprise command and control organizational structure, we identified supply, demand, and integrator nodes for enhanced ACS. In Table 6.1, we give examples of the roles and responsibilities that could be assigned to each node.

Air Force policy and instructions should detail the information flows into these nodes, processes that take place within the nodes, and products that leave one node for others. A defined organizational structure promotes clearly delineated roles and responsibilities, process activities, and information flows assigned to each node. As contingencies evolve, specific organizations can be designated to fulfill the responsibilities of each node. There might be variation in organizational construct by theater (nomenclature only), but roles and responsibilities should be standardized across AORs. And having standing command and control nodes could enable continuous combat support planning, execution, monitoring, and control for ongoing contingency and training operations worldwide.

There are many different options for how to assign these nodal responsibilities. ACS planning, execution, monitoring, and control could be managed theater by theater with decisionmaking retained at the theater level (see the "Forward" column in Figure 6.3). Or ACS planning, execution, monitoring, and control could be managed at the theater level with some reliance on outside organizations to provide resources and support (see the "Regional" column in Figure 6.3). Or, finally, ACS could be managed as a global enterprise supporting all ACS requirements (see the "Global" column in Figure 6.3).

We use Figure 6.3 to illustrate where ACS planning, execution, monitoring, and control processes could be conducted (forward, region-

Table 6.1
Defining Agile Combat Support Planning, Execution, Monitoring, and Control Processes to Supply, Demand, and Integrator Nodes

Node	Combat Support Planning, Execution, Monitoring, and Control Processes
Demand side	
COMAFFOR	Plan and assess support for air campaign – Review OPLAN/mission requirements – Identify combat support resource requirements/timing – Configure combat support resource infrastructure (direct intratheater/lateral support) – Assess support feasibility – Monitor/evaluate performance Plan and assess force beddown – Conduct site surveys – Plan for beddown – Assess plan feasibility – Project sustainment demands Plan and assess sustainment – Identify distribution nodes – Estimate movement requirements – Plan transportation routes/timing – Monitor/evaluate performance
Supply side	
Operational and support unit force providers	Plan/assess unit readiness and capabilities – Identify unit readiness requirements – Assess force and support capabilities – Monitor and resource shortfalls
Nonunit resource providers	Plan/assess enterprise supply-side plans – Develop/evaluate supply-side plans to meet demands developed by COMAFFORs – Assess resource capabilities – Assess resource levels – Assess resource surge capacities and needs
Integrator	
Air Force integrator	Monitor and assess global resource allocations – Integrate multitheater requirements – Identify capacities by commodity – Conduct integrated assessments (sortie production and base support) – Recommend allocation actions for critical resources

Figure 6.3
Options for How to Assign Agile Combat Support Planning, Execution, Monitoring, and Control Nodal Responsibilities

Environment	Process alignment		
	Forward	Regional	Global
Known large scale threats in a few AORs (forward model)	• Host nation support management • ATO support: sortie capability assessments • Site survey/beddown plan • Force-closure analysis • Theater distribution plans • Supply/demand arbitration within AEFTF	• Liaison with USTRANSCOM on strategic distribution • Sustainment planning	• Monitor contingencies • Resolve combat support shortfalls • Global assessments
Multiple contingency possibilities within AORs (regional model)	• Host nation support management • ATO support: sortie capability assessments	• Liaison with USTRANSCOM • Sustainment planning • Site survey/beddown plan • Force-closure analysis • Theater distribution plans • Supply/demand arbitration between AEFTFs	• Monitor contingencies • Resolve combat support shortfalls • Global assessments • AEFTF capability assessments
Multiple contingency possibilities within and across multiple AORs (global model)	• Host nation support management • ATO support: sortie capability assessments	• Site survey/beddown plans • Supply/demand arbitration between AEFTFs	• Monitor contingencies • Resolve combat support shortfalls • Global assessments • Liaison with USTRANSCOM on global distribution • Force-closure analysis • Sustainment planning • Supply/demand arbitration between AORs

(Right margin arrow, top to bottom: Limited resources and country access)

(Bottom arrow, left to right: System advances)

NOTES: AEFTF = Air and Space Expeditionary Force (AEF) Task Force. We show different activities in different colors so the reader can track how those activities shift as operations move from a large-scale threat in a few AORs (forward model) to multiple contingencies across multiple AORs (global model). For example, site surveys are conducted forward in the forward model, but they might be conducted regionally in the global model.

RAND *MG1070-6.3*

ally, or globally), depending on the type and location of the threat (from known in a single location to multiple threats across AORs). As technologies advance and become more reliable, enhanced ACS processes and organizational responsibilities can become more global. Personnel might not need face-to-face interaction at an FOL if authoritative data are available worldwide. Management of scarce resources can be conducted at a central location and pushed to the users.

From a resource viewpoint, likely continued budgetary pressures for efficient operations drive a global ACS construct. To respond to changing threats, combat support resources need to be reallocated from one theater to another. Currently, some resources are confined to individual theaters and are managed by theater-based organizations; others are managed by units. Both constructs make it difficult to relocate or reallocate resources to other AORs. If resources need to be allocated across competing AORs, they need to be managed from a global perspective—to enable moving limited capabilities quickly from one theater to another—with prioritization coming from a global combat support asset or functional manager.

As discussed earlier, the combat support community has moved to create some of these global supply chains and ACS functional managers. The GACP manages the munitions supply chain, the AFGLSC manages spare parts, and AFCESA manages civil engineering UTCs (personnel and equipment). However, these organizations have stovepiped resource orientations, and currently no organization is appointed to conduct integrated capability assessments (for example, integrating munitions with spares and civil engineering UTCs).

Management and control of individual materiel resources need to be integrated with other categories of materiel, including POL and personal equipment, to determine how all materiel interrelates in terms of affecting operational objectives or capabilities. Modeling and analysis capabilities, as described in the previous chapter, are needed to relate ACS resource levels and process performances to operational capabilities. These analyses would provide insights on how to allocate scarce resources among competing demands. Ultimately, the goal should be to determine how alternative resource allocations affect bombs on target or other desired effects. In the meantime, several operationally relevant metrics, such as the ability to generate desired missions, the ability to establish and sustain the desired number of FOLs, the ability to provide required security, or the ability to evacuate specific numbers of wounded or sick, can help guide allocations of scarce resources. The analysis of these metrics provides meaningful data to operations planners for any necessary replanning that ACS constraints might require.

Each capability assessment would include materiel, personnel, and infrastructure, along with strategic and tactical transportation capabilities. We represent the individual supply chains (storage, maintenance, procurement, and movement) on the bottom right side of Figure 6.4. The individual supply chains need to be integrated with other ACS functional capabilities, such as civil engineers, security forces, medical services, and maintenance, to give an overall picture of ACS capability. Assessments that integrate across supply chains and ACS functional capabilities, and others, such as transportation and infrastructure, are critical to the management and control of all resources necessary to initiate and sustain operations in both contingency and training environments (also shown on the right side of Figure 6.4 above the individual supply chains and ACS functional stovepipes).

A range of options exists regarding the assignment of global combat support integrated assessments and integrated control responsibilities across ACS capabilities. Three primary alternatives are listed here:

- A single organization could be given global responsibility for both integrated assessments and integrated control across the ACS capabilities necessary to produce combat support effects (see Figure 6.5).
- A single organization could be given global responsibility for conducting integrated assessments, identifying the most-binding constraints, requesting mitigation strategies for those constraints, providing assessments to higher authorities, executing resource allocation strategies when notified by higher authorities, monitoring actual ACS resource levels and performance against those authorized, and notifying appropriate organizations when in-control limits are breached and replanning must take place. In

Figure 6.4
Some Key Agile Combat Support Planning, Execution, Monitoring, and Control Processes Have Not Been Fully Developed or Assigned to Organizations

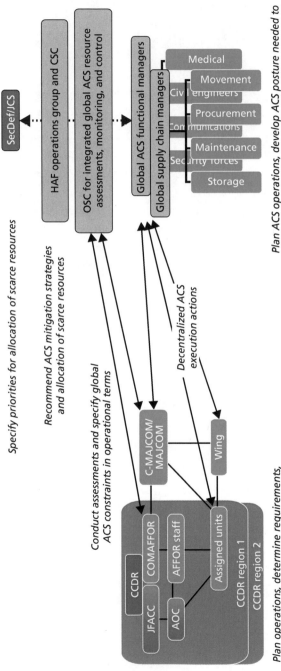

NOTE: Joint organizations are shown in purple, Air Force organizations are shown in blue, and processes that we defined in this analysis that are not currently assigned to an organization are shown in orange.

RAND *MG1070-6.4*

this option, separate supply chain and ACS functional managers could retain global control of assets across each ACS capability and execute to approved resource allocation plans (see Figure 6.6).

- Separate supply chain and ACS functional managers could retain both the capability to perform resource assessments and global control of assets, with integration of capability assessments being conducted when deemed necessary by the advent of specific contingencies or as issues arise in supporting ongoing training operations by an Air Force ACS GIC (see Figure 6.7).[2]

Alternatives 1 and 2 would establish standing assessment organizations that would continually monitor the capabilities of the global combat support system. Alternative 3 would not establish a standing organization to conduct integrated assessments routinely and would instead generate integrated assessments on an ad hoc basis, as contingencies or exercises require. We examine each alternative in turn.

Alternative 1

There are several options for where an independent agency to conduct integrated assessment and integrated command and control could be located. The AFGLSC has already been established as the global manager for spare parts. Its scope could expand to include global management responsibilities for other categories of materiel, such as end items (pods and engines), which would then be integrated with other combat support capabilities, such as personnel and lift, necessary to produce combat support capabilities.

Or, an OSC could be collocated with the AOC OSF and COMAFFOR reachback staff at the Ryan Center at Langley AFB in Virginia. Since the AOC OSF provides C-NAF reachback capa-

---

[2]   During Joint Expeditionary Force Experiment 11-1 in January 2011, the Air Force Command and Control Integration Center (AFC2IC) conducted an experiment called the Agile Logistics Evaluation eXperiment (ALEX). In ALEX, the AFC2IC stood up an ACS cell in the operational support facility (OSF) at the Ryan Center (Langley AFB, Virginia) to conduct integrated ACS assessments for existing OPLANs. This experiment demonstrated the GIC concept as outlined in alternatives 2 and 3 above. The GIC concept will be tested again in another ALEX in August 2011.

Figure 6.5
A Single Organization Responsible for Integrated Assessments and Resource Control

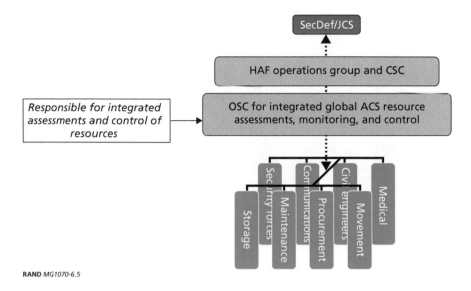

RAND *MG1070-6.5*

bility, personnel in this facility should already possess some of the knowledge and skill sets that would be useful in conducting global assessments—planning experience, including COA development and resource expertise.

Other options include collocating an OSC with the Directorate of Air and Space Expeditionary Force Operations at the Air Force Personnel Center (AFPC), which used to be the Air Expeditionary Force Center at Langley AFB in Virginia; at the Air Staff; or at a sourcing command, such as ACC. Or, a field operating agency (FOA) or a direct reporting unit (DRU) could be created to conduct these integrated assessments. A FOA would report directly to a HAF functional manager similar to the way in which the Air Force Agency for Modeling and Simulation reports to the SAF/XCC. A DRU would report directly to the Chief of Staff of the Air Force (CSAF) (or a designated representative at HAF) in the same manner in which the Air Force Academy operates. Each option offers benefits, but each should be evaluated for associated costs and risks before a decision is made on agency location.

Alternative 2

Under this alternative, supply chain and ACS functional managers for individual resources would retain global control over assets. The AFGLSC would retain its current scope focused on spare parts; ACC Plans and Integration would be the global WRM manager, with positioning and sourcing responsibilities for WRM; and AFCESA would continue to monitor and control civil engineering UTCs. Separate assessment centers for materiel, personnel, infrastructure, and lift would monitor and evaluate how each category of capability comes together to produce combat support capabilities for the warfighter. (Note that, under alternative 1, the individual assessment centers would exist within the single organization chosen for integrated assessment, monitoring, and control.)

Figure 6.6
An Organization Responsible for Integrated Assessments but Not Resource Control

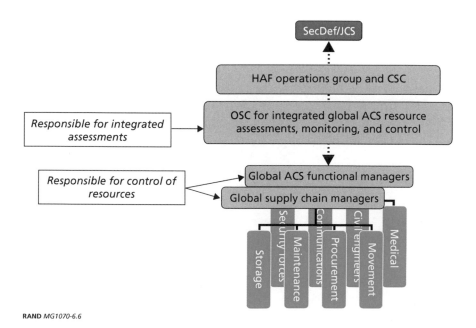

Alternative 3

Here again, supply chain and ACS functional managers for individual capabilities would retain global control over assets; however, an individual assessment-center capability would be established to perform integration analysis as required on a scheduled or an ad hoc basis. This alternative would not involve assessments on a continuous basis but would have them conducted when scheduled—for example, for LSAs or exercises.

Under all options, individual assessment-center responsibilities would need to be integrated with the other pillars of combat support sustainment analyses—that is, materiel, infrastructure, combat support forces, and lift as outlined in AFI 10-401—to make resource trade-offs and adjudicate cross-CCDR allocations in support of contingency operations or cross-MAJCOM allocations to meet training responsibilities. These individual combat support assessment centers

Figure 6.7
Integrated Assessments Conducted Only When Needed Without Resource Control

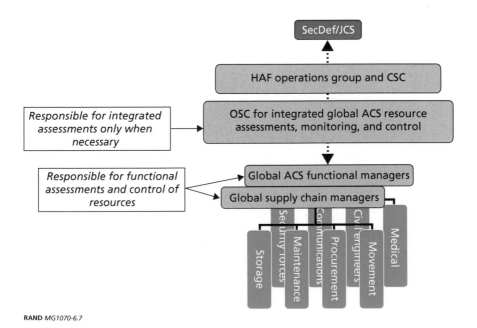

could reside at the Air Staff so as to have the appropriate level of visibility across resources and capability demands.

All three options would provide enhanced enterprise-level ACS planning, execution, monitoring, and control and thereby improve Air Force and joint command and control. Each alternative would better evaluate options for integrating resources to achieve specified operational objectives. In this way, the GIC acts as reachback support to forward C-NAF staff personnel to evaluate the supportability of different options for combining different resources to achieve specified objectives.

Under alternative 1, the GIC would have the authority to make asset and personnel positioning and sourcing decisions, in both the contingency planning and execution spaces, based on global priorities. This alternative would also have the GIC staffed permanently.

Under alternative 2, the GIC would conduct assessment and control functions and direct supply chain managers and FAMs of approved resource allocation decisions. The individual supply chain managers and FAMs would carry out the instructions and notify the GIC when their activities breach approved process performance, personnel, or materiel assets' control levels. This alternative would also have the GIC staffed permanently.

Under alternative 3, the GIC would conduct assessments on a scheduled or ad hoc basis as needed. Note that, under all alternatives, each ACS resource assessment, whether materiel or human, is managed and controlled by a single global node, which orchestrates physically distributed ACS supply chain managers and FAMs and connects them virtually.

For example, for WRM, the GIC would conduct the necessary WRM analyses to determine (or suggest) what positioning and sourcing decisions could satisfy CCDR needs as specified by the COMAFFOR logistics and installation personnel. The GIC would house the tool sets necessary to identify supportable delivery timelines and risks to CCDRs, the Air Staff, and the Joint Staff.[3]

However, WRM needs to be integrated with other materiel categories in evaluating operationally relevant MOEs. For example, consider the spare parts, engines, and other commodities needed to support deploying F-16 units. The AFGLSC has the responsibility to work with the units to determine RSP needs, which is a function of how long each unit will be deployed and how much flying each unit is preparing to accomplish. If the units have lengthy deployment times projected with relatively high activity levels, an in-theater CCRF might need to be established to perform phase inspections or to repair end items or aircraft spares. The global WRM manager, whether integrated with other commodities or not, would need to work with the COMAFFOR logistics and installation staff and the AFGLSC not only to identify the WRM movements necessary to support establishment of an initial operating capability when needed at the FOLs but also to work the timelines to establish CCRFs, and the AFGLSC would work the bench stock and associated component spares to support the CCRF. This analysis of closely coupled materiel options is similar to what the AOC does for the various operational platforms.

Alternative 1 would likely be the most expensive in terms of personnel. In this option, the GIC, operating continuously, would house most of the personnel charged with executing supply chain instructions and moving personnel from one UTC to another to robust a given set of UTCs. Alternative 2 would require a relatively small staff to maintain liaison with C-NAFs, supply chain managers, and FAMs and to run assessments as needed. Alternative 3 is similar to alternative 2 but would leverage ANG and AFRC capabilities to attract and retain the needed staff to conduct ACS assessment and control function responsibilities. This option is likely the least expensive one.

In all of the options, the GIC organization would need significant investments in modeling capabilities and staff development to allow it

---

[3]   We note that this type of assessment was conducted before OIF by an ad hoc group established for this purpose by AF/A4/7 working with the AFCENT Logistics Directorate (AFCENT/A4).

to perform risk assessments within the short decision cycles required by military leadership. A critical aspect of this global management concept is the determination of the guaranteed levels of support to CCDRs mentioned earlier. The global manager would need to communicate regularly with CCDRs' staff to help assess the feasibility of modifications to scenario plans under current funding levels and to understand each commander's priorities across various scenarios. Then, provided that the global asset managers can sustain these guaranteed performance levels, day-to-day management activities could occur entirely within the global manager's purview: Elevation of decisions to higher authorities would be necessary only when the state of the combat support system varied so far from plans as to make the guaranteed levels of support to every CCDR no longer sustainable.

Agile Combat Support Planning, Execution, Monitoring, and Control Organizational Structure Summary

The preceding discussion raises big issues, such as how ACS planning, execution, monitoring, and control nodes might be best organized to carry out their command and control functions. A strategies-to-tasks analytic approach could be useful in examining the pros, cons, costs, and benefits of options for meeting GIC and integrated analysis responsibilities. Currently, decisions are pending on where to place the GIC function. In addition, the Air Force has taken an enterprise view of the repair enterprise, and issues about how to measure enterprise repair capacity and capability exist, as do issues of who should manage and shape the repair enterprise of the future. This is a large responsibility and is an important "brain–command and control" function—as is the AFGLSC. Options for managing and controlling the development of the repair enterprise need to be thought through, and the modified strategies-to-tasks framework can be useful in evaluating options for this activity as well.

Combat support resource assessment and allocation management could be assigned to permanent organizational nodes dedicated to resource monitoring, prioritization, and reallocation. Additionally, having a standing integration function for combat support resource management could facilitate the incorporation of relevant data into capability assessments and raise these assessments' visibility in the eyes of the operational community. Regardless of the organizational structure adopted, the roles and responsibilities of each ACS organizational node, as well as each node's interaction with joint combat support nodes, should be clearly defined and documented in Air Force doctrine and guidance, including information needed, processes, and information produced at each node.

CHAPTER SEVEN

Conclusions and Recommendations

Conclusions

Creating a process, clearly defined in doctrine, to specify ACS planning, execution, monitoring, and control supply, demand, and integrator roles, including what information flows, in what format, and to whom, could lead to better integration between combat support and operations. As part of the ACS planning, execution, monitoring, and control processes, the combat support community should be able to relate combat support resources and process performance to operational effects. Combat support personnel might need to continue to monitor each piece and pipeline within the system, but combat support parameters should be synthesized into metrics that are well understood by the operational community, such as sortie generation capability and FOL IOC.

In addition, developing a closed-loop planning and execution process, acting within operational decision timelines, with established control parameters against which to track actual combat support performance and signal when a process' performance exceeds or falls short of objectives could aid in making ACS more proactive rather than reactive to changing operational requirements. This too could lead to better coordination, timeliness, and accuracy of combat support planning and added value of ACS to the operational community.

The absence of well-defined supply, demand, and integrator processes, delineated in policy, contributes to a shortfall in training. Many ACS personnel do not understand how to apply the nonmarket, resource-constrained strategies-to-tasks and closed-loop frameworks to maximize efficiency and effectiveness. More training and expanded educational opportunities are needed on relating combat support options to the CCDR's campaign plan to achieve joint operational effects.

Current information and tool shortfalls reflect current process shortfalls. Decision-support tools and job-performance aids should complement formal courses and exercises. Existing Air Force systems and prototype tools can be leveraged to provide enhanced information and data for ACS planners; however, new tools might need to be developed to provide an integrated view of combat support resource allocations and process performance. Properly integrated information could greatly reduce the risk of operational failure, the need to revise plans midstream, allow a faster transition to operations and better-informed decisions, and facilitate adjustments when necessary.

And finally, global management and control of combat support capabilities could facilitate resource allocation assessments among competing CCDRs to inform tough capability trade-off decisions. These assessments should be used to inform POM and other budgeting and program decisions. However, with global management comes some risk of single-point failure. Methods to provide COOP and to minimize network vulnerabilities need to be developed.

Recommendations

In this monograph, we discuss options for improving ACS planning, execution, monitoring, and control. Some are short-term solutions with little implementation cost. Other improvements will take time, resources, planning, and programming. Cost estimates can be developed for options that senior Air Force leaders view as high priority for implementation. Table 7.1 summarizes our recommended actions to enhance ACS planning, execution, monitoring, and control within Air Force enterprise command and control.

Although Air Force transformational initiatives (both operational and in combat support) have moved the Air Force forward in achieving the enhanced ACS TO-BE vision, our research shows that many

Table 7.1
Recommended Actions for Improved Agile Combat Support Planning, Execution, Monitoring, and Control Within Enterprise Air Force Command and Control

Goal	Action Required to Achieve the Goal
Enhance processes	Focus ACS planning, execution, monitoring, and control processes on operational outcomes; identify and separate supply, demand, and integrator processes; include closed-loop feedback and control
Expand doctrine	Delineate roles of ACS nodes, including logistics, operational, and installation staff; Air Force commanders; MAJCOMs; the AFGLSC; and others
Refine training and expand education	Educate Air Force staff officers in ACS planning and staff responsibilities and strategies-to-tasks methodology; assign some promotable supply-side officers to demand-side organizations and vice versa
Implement systems and tools	Identify critical ACS communications and information-system capabilities needed to assess, monitor, and inform allocation decisions, and update as necessary
Strengthen organizations and instructions	Assign supply, demand, and integrator processes to organizations and functions; modify instructions and other documents to support ACS assessment and control functions

actions still can be taken to improve ACS planning, execution, monitoring, and control processes, doctrine, training, tools, and systems. A key aspect of this is updating the 2002/2006 enterprise command and control OA to reflect the current operational environment. Using this analysis, RAND researchers worked with the Air Force to perform a comprehensive review of all the combat support functional capabilities, as identified in the ACS CFMP, and updated the enterprise command and control OA.[1] We focused our efforts on nodal roles and responsi-

---

[1]   One forthcoming PAF document describes in detail a strategic- and operational-level C2 architecture integrating enhanced ACS processes. A second analysis documents how we used the architecture to identify and describe where shortfalls or major gaps exist between current ACS processes (the AS-IS) and the vision for integrating enhanced ACS processes into Air Force C2 (the TO-BE).

bilities to provide an incremental approach of how ACS planning, execution, monitoring, and control processes can be incorporated within the Air Force command and control enterprise.

The RAND Strategies-to-Tasks Framework

The RAND-developed strategies-to-tasks framework links strategic national goals to operational tasks (see Figure A.1). The framework was designed to aid in strategy development, campaign analysis, and modernization planning.[1] It has proven to be a useful approach for providing intellectual structure to ill-defined or complex problems. If used correctly, it links resources to specific military tasks that require resources, which, in turn, are linked hierarchically to higher-level operational and national security objectives. Working through the strategies-to-tasks hierarchy can help identify areas in which new capabilities are needed, clarify responsibilities among actors contributing to accomplishing a task or an objective, and place into a common framework the contributions of multiple entities and organizations working to achieve some common objective.

At the highest levels of the strategies-to-tasks hierarchy are *national goals*, which are derived from U.S. heritage and are embodied in the U.S. Constitution. These national goals do not change over time. They form the foundation from which all U.S. statements regarding national security are derived.

The *National Security Strategy* is formulated in the executive branch. It outlines strategy for applying the national instruments of power—political, economic, military, and diplomatic—to achieve U.S. national security objectives.

---

[1] Internal examples are Lewis, Coggin, and Roll, 1994, and Niblack, Szayna, and Bordeaux, 1996. Outside of RAND, the framework is in use by the Air Force, the Army, and elements of the Joint Staff.

Figure A.1
Strategies-to-Tasks Hierarchy

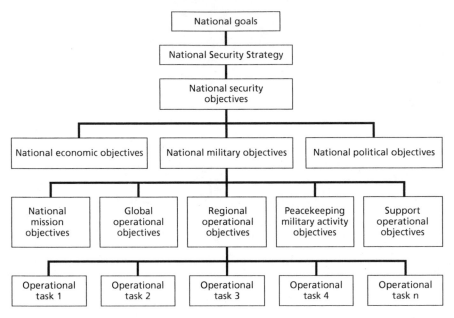

RAND *MG1070-A.1*

National security objectives define what must be done to preserve and protect fundamental U.S. goals and interests from threats and challenges that originate abroad. In contrast with national goals, national security objectives change in accordance with changes in the geopolitical environment. For example, the national security objectives specified in *The National Security Strategy of the United States* states that the United States must do the following (Office of the President of the United States of America, 2006, p. 1):

- Champion aspirations for human dignity.
- Strengthen alliances to defeat global terrorism and work to prevent attacks against the United States and its friends.
- Work with others to defuse regional conflicts.
- Prevent the nation's enemies from threatening the United States, its allies, and its friends with weapons of mass destruction.

- Ignite a new era of global economic growth through free markets and free trade.
- Expand the circle of development by opening societies and building the infrastructure of democracy.
- Develop agendas for cooperative action with other main centers of global power.
- Transform U.S. national security institutions to meet the challenges and opportunities of the 21st century.
- Engage the opportunities and confront the challenges of globalization.

National military objectives are formulated by the SecDef and CJCS. The national military objectives define how the United States will apply military power to attain national objectives to support the National Security Strategy. Collectively, they define the National Military Strategy (NMS), which identifies (at a high level) how the United States will respond to threats to its national security. For example, as defined in the *National Military Strategy of the United States of America*, these are to do the following (JCS, 2004):

- Protect the United States against external attacks and aggression.
- Prevent conflict and surprise attack.
- Prevail against adversaries.

Operational objectives describe how forces will be used to support the national military objectives. They can be regional or global and include support activities necessary to sustain military operations. For Operation DESERT STORM, an example of a political objective might be to expel Saddam Hussein from Kuwait. The regional operational objective was to cut off communications and destroy supply lines. To accomplish the objective, the air component was tasked to maintain air superiority.

Tasks, formulated by the CCDRs and their staffs, are the specific functions that must be performed to accomplish an operational objective. Operational tasks constitute the building blocks of the application of military power.

The Strategies-to-Tasks Framework Can Help Identify Operationally Relevant Metrics

The operational community uses an approach similar to the RAND strategies-to-tasks framework in its UJTL and COA development processes to demonstrate operational effects. It outlines how national goals can be disaggregated into national diplomatic, economic, informational, and military objectives and how regional military operational objectives can be formulated from national military objectives. Joint operational tasks can be assigned to JTFs within the region.[2] Task-organized operational elements carry out the tasks assigned to them, and task-organized combat support elements provide the needed resources to conduct the operational mission.

The combat support community is not as skilled when it comes to relating resources to their operational effects. For example, if preventing conflict and surprise attacks is the national military objective (center of Figure A.2), the regional objectives could be to establish a presence in the country and engage the combatants (one level below the military objective in Figure A.2). The combat support tasks to achieve those objectives would be to beddown forces, deliver munitions, and generate sorties (bottom row of Figure A.2).

To beddown the forces, BEAR assets might be needed to open an FOL to sustained operations in the field. Without the BEAR (the resource), the FOL cannot be opened, so the operational task cannot be accomplished (operational effect). Applying the strategies-to-tasks framework could help combat support planners better articulate the relationship between combat support resources to their operational effects.

Once combat support resources are better linked to operational effects, combat support planners should be able to better communicate combat support options in terms of metrics understood and appreciated by the operational community, such as sortie generation capability or FOL IOC. We found that most other combat support metrics can be rolled up and shown to affect one of those two metrics.

---

[2] The number and nature of these joint operational tasks will change over time.

Figure A.2
Using the Strategies-to-Tasks Framework to Identify Combat Support Tasks

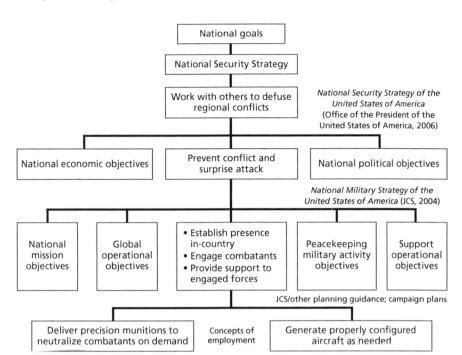

RAND *MG1070-A.2*

Adjudicating Requirements Within Economic Constraints

Linking national goals to operational tasks as outlined in the strategies-to-tasks framework is a necessary piece of the command and control system; however, the military does not operate in an open-market environment in which there is an unlimited supply of resources. Because of resource constraints, trade-off decisions are often necessary to prioritize requirements.

To address such trade-off decisions, we expand the strategies-to-tasks framework to include economic constraints, highlighting the need for resource allocation decisionmaking strategies. Resource allocation for ACS planning, execution, monitoring, and control can be viewed as a problem of integrating the *demand* for combat support resources with the available *supply* of combat support resources. Finally, we identify *integrator* processes, in this case, for accomplishing combat support objectives. Integrator processes are those processes associated with allocating scarce combat support resources according to prioritized CCDRs' needs. Figure A.3 provides an illustration of how resource allocation considerations can be integrated into a strategies-to-tasks framework that manages ACS planning, execution, monitoring, and control.

The left side of Figure A.3 lists some illustrative operational tasks (these are the tasks identified in the strategies-to-tasks framework, like the one shown in the bottom row of Figure A.2). We call this the demand side. The right side of the figure lists some illustrative force and support elements that can be selected from component providers

Figure A.3
Strategies-to-Tasks Framework with Resource Allocation Considerations

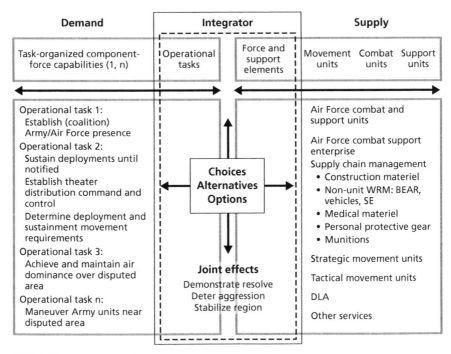

NOTE: SE = support equipment.

RAND *MG1070-A.3*

to satisfy the operational tasks on the left of the figure. We call this the *supply* side. The middle of the figure shows the integration of demand and supply processes. Here, the *integrator* chooses the force and support elements from the available options, each of which could have differing attributes and differing operational effects.[3] The result creates the combat support capabilities shown at the bottom of the "Integrator" box.

In this case, combat support choices can be made from a set of options that include service-provided assets or those available from other DoD or government sources. Each choice could result in differing capabilities—for example, different timelines for establishing coalition or joint Army and Air Force presence in the area of interest.

Each of the combat support operational tasks might require combinations of component resources to achieve the desired capability and, ultimately, the joint operational effect. For example, a CCDR might need assured, scheduled movement in several parts of the AOR. There might be more than one way to meet that demand—component assets or commercial assets. The integrator would make the decision about how to best meet all or part of the demand for assured movement using the spectrum of resources available. Figure A.4 shows a high-level representation of the expanded nonmarket, resource-constrained strategies-to-tasks framework that relates demand for combat support capabilities to supplies of combat support capabilities.

Using the Nonmarket, Resource-Constrained Strategies-to-Tasks Framework to Identify Process Roles and Responsibilities

CCDRs and their joint staffs formulate operational requirements based on the forecast missions for their theater. OSD guidance lays out the operational demands the CCDR should be prepared to support. How-

---

[3] It is the integrator's job to arbitrate between the demand and supply sides. To effectively arbitrate, the integrator needs capability assessments to make informed trade-off decisions. Without these assessments, the integrator has limited visibility into the effects of his or her trade-off decisions. Capability assessments are key to making informed integrator decisions.

Figure A.4
Nonmarket, Resource-Constrained Strategies-to-Tasks Framework for Agile Combat Support Planning, Execution, Monitoring, and Control Responsibilities

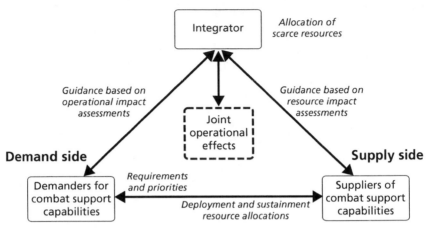

RAND *MG1070-A.4*

ever, the Air Force does not program for projected operational tempo for all operations (there are not enough assets to satisfy every requirement in every AOR). Thus, there are automatic shortages that need to be addressed. Assets must be allocated across competing demands according to SecDef priority. For the foreseeable future, the Air Force will continue to operate within a resource-constrained environment in which demand exceeds supply, highlighting the need for resource allocation decisionmaking strategies.

We use the nonmarket, resource-constrained strategies-to-tasks framework to identify processes associated with allocating scarce resources against prioritized CCDRs' requirements for accomplishing combat support objectives—that is, combat support supply, demand, and integrator roles and responsibilities. Figure A.5 shows a high-level strategies-to-tasks view of these process roles and responsibilities.

In this representation, the national command authorities (NCA), the JCS, and the SecDef act as integrators providing integrated guidance on the apportionment of forces, which results in the engagement of combat support capabilities. In this view, each CCDR is responsible

Figure A.5
Using the Strategies-to-Tasks Framework, the Secretary of Defense Is the
Neutral Integrator at the Highest Level

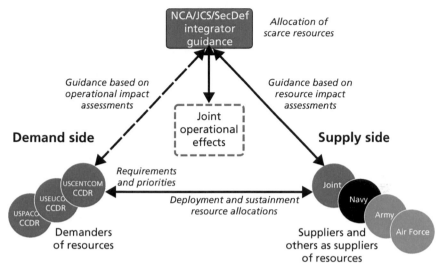

RAND *MG1070-A.5*

for estimating and prioritizing his or her combat support needs and requirements. These requirements could be gathered and prioritized by a Joint Staff Operations/Strategic Plans and Policy (J-3/5) organization through an integrated closed-loop process with the J-4 community, on the demand side.[4]

Also, from this view, the Air Force is a supply-side organization responsible for providing combat support capabilities for use by the CCDRs. These responsibilities include configuring combat support resources to meet needs, transmitting needed information to users, establishing schedules for meeting user needs, and overseeing execution operations. As illustrated in the figure, there are other providers of combat support capabilities, such as the other service components, and the integrator can choose who should supply the needed capabilities.

---

[4]   See Tripp, Lynch, Roll, et al. (2006) for TDS analysis that outlines proposed J-3/5 organizational roles and responsibilities.

A feature of supply-and-demand relationships is that they are often nested, both within and outside the command or service. Not only are supply, demand, and integrator roles defined at the execution level (as in Figure A.5); they exist at other levels as well. An organization can be a demand-side organization at one level and a supply-side organization at another level. For example, in Figure A.6, we show the COMAFFOR as a demand-side organization. Here, the COMAFFOR is a demander of Air Force combat support capabilities from suppliers, such as Air Force MAJCOMs—ACC, AMC, and AFMC—on behalf of the CCDR. USJFCOM and the Air Staff, with the JCS and the SecDef, act as integrators at this level.[5]

Figure A.6
The Commander, Air Force Forces, Is a Demander of Combat Support Resources at This Level

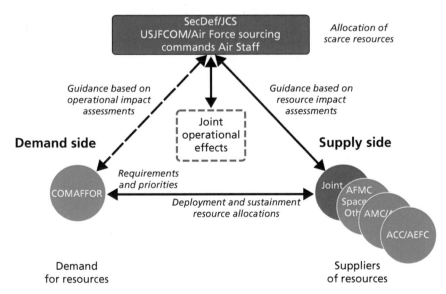

NOTE: AEFC = Aerospace Expeditionary Force Center.

RAND MG1070-A.6

---

[5] Recent DoD plans call for the closing of USJFCOM. When the command is dissolved, the neutral integrator processes currently conducted by USJFCOM will have to be reassigned to other organizations.

The COMAFFOR can also be a supply-side organization, along with the other services, supplying forces to a JTF (see Figure A.7). In this illustration, the JFACC would integrate across services to satisfy operational planning requirements.

The nesting that exists in planning and executing combat support processes adds another layer of complexity to the ACS system. An organization can serve in different capacities at different levels of responsibility. We use the nonmarket, resource-constrained strategies-to-tasks view to provide insights on combat support processes and assign processes among existing organizations. The process of stating the operational requirements (demand side), identifying the available resources to meet those demands (supply side), and arbitrating which demands will be met and how (integrator) is the first step in a closed-loop ACS system.

Figure A.7
The Commander, Air Force Forces, Is a Supplier of Combat Support Resources at This Level

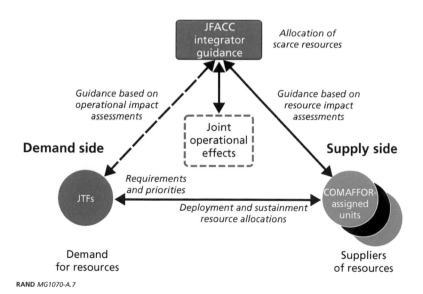

RAND *MG1070-A.7*

Agile Combat Support Annotated Bibliography

In this appendix, we list DoD, joint services, and Air Force publications reviewed as part of this ACS planning, execution, monitoring, and control analysis. For each publication, we list the title, date, and a synopsis of the relevant guidance as it pertains to ACS planning, execution, monitoring, and control. We point out where doctrine might be lacking, as well as where the guidance already supports enhanced ACS planning, execution, monitoring, and control processes we identify as necessary in the ACS vision of the future.

GEF, 2007

This DoD document is a capstone planning document and sets the stage and priorities for current force employment and planning across DoD. Of special note is its statement that DoD resource managers know that they need "a mechanism that can systematically identify all the demand signals from the field" but that there is "currently no system in place . . . that drives a comprehensive assessment of resource requirements [and] presents a complete picture to [DoD decisionmakers] for resource allocation and apportionment" (Chapter 5, Campaign Planning Construction, Section 2.e). It also states that plans "should use CONUS-based reach-back capabilities to the maximum extent possible" (Chapter 6, General Planning Requirements, Section 20). This is particularly important given the large, ongoing demand that current operations in Iraq and Afghanistan place on the services. The campaign plan will serve as the forcing function for identifying

COCOM-wide resource requirements and expose requirements that might not have been explicitly identified before—particularly for shaping and security cooperation purposes.

GFMIG, 2008

This document integrates force assignments, apportionment, and allocation processes to improve DoD's ability to manage forces from a global perspective. It provides force management implementation guidance and complements the GEF and the GDF. Global Force Management (GFM) goals are as follows:

- Account for forces and capabilities committed to ongoing operations and constantly changing unit availability.
- Identify the most-appropriate and most-responsive force or capability that best meets the COCOM requirement.
- Identify risks associated with sourcing recommendations.
- Improve the ability to win multiple overlapping conflicts.
- Improve responsiveness to unforeseen contingencies.
- Provide predictability for rotational force requirements.
- Identify forces and capabilities that are unsourced or hard to source (UHTS).

GFM enables the SecDef to make more-proactive, risk-informed force management decisions by integrating the three processes of assignment, apportionment, and allocation. These processes facilitate alignment of operational forces against known apportionment and allocation requirements in advance of planning and deployment preparation timelines. The end result is timely allocation of forces and capabilities necessary to execute CCDR missions (including theater security cooperation tasks), timely alignment of forces against future requirements, and informed SecDef decisions on the risk associated with allocation decisions.

DODD 7730.65, 2007

This directive establishes DRRS to measure and report readiness of military forces and the supporting infrastructure through the use of capability-based, adaptive, near–real-time readiness reporting. It applies to OSD, military departments, CJCS, CCDRs, defense agencies, field activities, and all other organizational DoD entities. All DoD components are directed to align readiness-reporting processes in accordance with this directive and use DRRS to identify critical readiness deficiencies, develop strategies for rectifying these, and ensure that they are addressed in program and budget planning and other DoD management systems. Components develop mission essential tasks (METs) for all assigned missions to collect information about readiness of military forces and support organizations to perform these missions. Within DRRS, Enhanced Status of Resources and Training System (ESORTS) captures metrics and supporting data on readiness. ESORTS highlights deficiencies in the areas of training, personnel, equipment, ordnance, and sustainment.

DRRS CONOPS, 2009

This CONOPS serves as the vision document for transforming the readiness-reporting construct of DoD and is approved by consensus of the DRRS Executive Committee. The overarching vision for DRRS is to both extend and expand the existing readiness reporting paradigm while using the collaborative capabilities of net-centric information technology (IT) systems. Readiness reporting will be extended via METs and overall mission assessments to provide a "capability-based" appraisal of unit and organizational readiness to accomplish specified tasks and missions. Readiness reporting will be expanded by complementing the traditional bands of resource and training data currently resident in Global Status of Resources and Training System (GSORTS) (overall C-ratings and associated P, S, R, and T levels), with "authoritative" data obtained by querying, organizing, and displaying the underlying data from various authoritative data sources.

DRRS will provide CCDRs and U.S. Special Operations Command (USSOCOM) with relevant readiness data, in the form of capability assessments supported by resource status to help determine whether they can perform their assigned missions and associated METs in a joint, interagency, and multinational operational environment. Involved with this determination are the service component assessments of their ability to conduct missions as part of a joint organization, according to the specified conditions and standards of the joint commander's capability and MET requirements. Equally, the services and combat support agencies (CSAs) will gain an unambiguous view of CCDR capability requirements in clear operational terms, i.e., through the JMETL and specified subordinate tasks.

A central tenet behind the DRRS vision is the importance of sharing data: By leveraging IT concepts, such as web-based services and service-oriented architecture, independent systems can integrate with DRRS to share information requirements and data elements seamlessly across the enterprise.

JP 0-2, 2001

Unified Action Armed Forces (UNAAF) provides the basic doctrine and policy governing the unified direction of forces and discusses the functions of DoD and its major components. It provides guidance for the exercise of authority by CCDRs and other joint force commanders (JFCs), prescribes doctrine for unified actions and joint operations and training, and provides military guidance in preparing appropriate plans. The UNAAF covers roles, missions, functions, composition, command relationships, joint command and control, multinational operations, and establishment of unified, specified, subordinate unified commands and JTFs.

These principles and guidance might also apply when significant forces of one service are attached to forces of another service or when significant forces of one service support forces of another service.

Directive authority for logistics by a CCDR includes the authority to issue directives to subordinate commanders, including peacetime measures necessary to ensure the following: effective execution of approved OPLANs, effectiveness and economy of operation, and prevention or elimination of unnecessary duplication of facilities and overlapping of functions among the service component commands. A CCDR's directive authority does not discontinue service responsibility for logistic support, discourage coordination by consultation and agreement, or disrupt effective procedures or efficient use of facilities or organizations.

JP 3-0, 2008

JP 3-0 is the keystone document of the joint operations series and provides doctrinal foundation and fundamental principles that guide armed forces in the conduct of joint operations across the range of military operations. It defines strategic context, fundamentals of joint operations, joint functions, planning, operational art and design, assessment (for the purpose of determining the joint force's progress toward mission accomplishment), major operations and campaigns, crisis response and limited contingency operations, and military engagement, security cooperation, and deterrence.

JP 4-0, 2008

This is a keystone document on joint logistics, concentrating on sustainment yet providing the doctrinal framework on how logistics is delivered to support joint operations. It describes the joint environment, key imperatives, integrating functions, roles, core logistics capabilities, planning and considerations, execution, and control mechanisms that enable synchronization of logistics in support of the CCDR. Since logistics remains a service responsibility, rarely will the joint logistician have unity of logistics command. Because of this fact, the CCDR has the authority to organize logistics resources within theater according to operational needs. This publication further provides functional relationships between the CCDR's J-4 staff and joint DDOC (JDDOC) at USTRANSCOM. Logistics command and control resides in the services at the operational level and logistics control structures for the basis for joint operations.

JP 4-0 states, "Control of joint logistics is reflected by how effectively the logistician combines the capabilities of the global providers and the requirements of the CCDR in a way that achieves unity of effort." Joint logisticians must integrate service, multinational, agency, and other organizational capabilities and resources to plan, execute, and control logistics in support of the CCDRs' CONOPS.

JP 5-0, 2006

This document provides joint doctrine to govern the joint operation planning activities and performance of the armed forces in joint operations, and it provides the joint doctrinal basis for coordination with other agencies and U.S. military involvement in multinational operations. It covers types and scope of joint strategic planning, organization and responsibilities, types of plans, joint operation planning and execution systems, availability of forces, global planning, strategic direction, national-level systems, the JOPP in detail, and operational art and design. Assessment is discussed, but, as with JP 3-0, its definition and scope focus on plan execution assessment and how to build appropriate MOEs into the plan. It does not describe force readiness in preparation for support to the CCDR. Sustainment is discussed as an assured capability to be addressed in each planning event and where in each type of plan it should be positioned. Command and control issues with logistics are not specifically addressed.

JP 4-01.4, 2000

This publication defines joint TTP for theater distribution in joint operations for a CCDR and staff. It provides theater distribution concepts and governs the joint theater distribution activities. It reviews the roles and responsibilities of the many individuals and organizations involved in joint theater distribution. Additionally, it provides theater distribution planning and operational considerations and defines the joint communications and information systems utilized in theater distribution. Regarding ACS planning, execution, monitoring, and control, the publication states the following:

> Predeployment planning is an integral part of logistics preparation of the theater C2 capability and requirements and C2 requirements and responsibilities for logistics. Each service is designated to provide logistics support in accordance with their executive agent responsibilities, combatant commander–directed service responsibilities, Title 10, OPLANs, and OPORD [operations order]–designated responsibilities. Great latitude is given to the CCDR on how to organize theater distribution.

JP 3-35, 2007

This publication provides doctrine and principles for planning and executing deployment; joint reception, staging, onward movement, and integration (JRSOI); and redeployment. It explains the deployment, JRSOI, and redeployment processes and planning and execution considerations that could affect force-projection operations. It discusses the responsibilities and command relationships for supported and supporting COCOMs and services and the interaction with other DoD and federal agencies, foreign nations, allies, multinational organizations, and other groups.

Sustainment is addressed as a process delivery—specifically, as follows: Sustainment delivery is the process of providing and maintaining levels of personnel and materiel required to sustain combat and mission activity at the level of intensity dictated by the CONOPS. Sustainment is ongoing throughout the entire operation and, like deployment and redeployment, should be aligned with the mission and mission priorities of each phase. Sustainment delivery must frequently be balanced against force deployment or redeployment requirements because these operations share the same deployment and distribution infrastructure and other resources. However, deployment and force integration can be adversely affected by excess or insufficient sustainment support; hence, operation planning must integrate deployment and sustainment operations.

Also of note are USJFCOM's responsibilities: USJFCOM serves as an integrator of capabilities from the five services, Reserve Component (RC), and interagency sources. USJFCOM's integration effort is focused primarily on developing and maintaining technological interoperability among service systems employed by joint headquarters and staffs. USJFCOM also serves as the JFP for conventional forces. Supported by its four service component commands (U.S. Army Forces Command, U.S. Marine Corps Forces Command, U.S. Fleet Forces Command, and ACC), USJFCOM identifies conventional-force sourcing solutions in response to supported CCDR requirements.

Specific Air Force responsibilities are defined as follows: The Air Force relies on common-user transportation to move support forces and sustainment cargo. Within the AFFOR component, the A4 is the principal coordinator of Air Force logistics. When required, the A4 provides centralized direction and control of deployment, reception, integration, employment, and redeployment of logistics and support assets.

JP 5-00.2, 1999

This publication provides fundamental guidance and procedures for the formation and employment of a JTF to command and control joint operations throughout the range of military operations. It details JTF organization and staffing, subordinate commands, JTF command and control, manpower and personnel, intelligence, JTF operations, JTF logistics, JTF plans and policy, and command, control, communications, and computer systems.

The JTF J-4 is charged with the formulation of logistics plans and with the coordination and supervision of supply, maintenance, repair, evacuation, transportation, engineering, salvage, procurement, mortuary affairs, security assistance, host nation support, and related logistics activities.

It is critical that the JTF J-4 determine what, if any, logistics directive authority for a common support capability the supported CCDR has delegated to the combined (or coalition) joint task force and whether the scope of the authority meets the JTF requirements. The JTF J-4 can establish a logistics readiness center (LRC) to coordinate logistics support, maintain total asset capability, monitor logistics capability, and provide a central point for logistics-related boards, offices, and centers.[1]

Logistics will play a key role in JTF operations from the earliest stage of planning through the final stage of redeployment of forces. The joint staff's J-4 organization should be tailored to the operation. The JTF J-4 should consider forming a JTF LRC and a joint movement center. A J-4 logistics staff representative must be included in all JTF planning, including permanent membership in the joint planning group (JPG). J-4 responsibilities and authority must be clearly delineated to ensure uninterrupted sustainability of ongoing and future operations.

CJCSI 3110.01G, 2008

This instruction provides guidance to CCDRs, service chiefs, CSA directors, applicable defense agencies and directors of DoD field activities, and the chief, NGB, to accomplish tasks and missions based on near-term military capabilities. The Joint Strategic Capabilities Plan (JSCP) implements campaign support, contingency, and posture planning reflected in the GEF. GEF guidance is not repeated in the JSCP. The GEF is a companion document to the JSCP for planning.

The JSCP provides strategic planning direction for campaign, campaign support, contingency, and posture planning to be developed; details planning guidance, force apportionment, assumptions, and tasks; tasks CCDRs to prepare plans and security cooperation

---

[1]   JDDOCs are not discussed or referred to because JP 5-00.2 predates their formation.

guidance; and establishes synchronizing, supported, and supporting relationships.

CJCSI 3110.03C, 2007

This document provides logistics planning guidance to the CCDRs, chiefs of the services, and heads of the DoD agencies in support of the tasks assigned in the JSCP. Within the Joint Strategic Planning System (JSPS), the basic purposes of logistics planning are to determine logistics requirements, establish logistics planning responsibilities, evaluate logistics capability to execute joint operations in support of the CCDRs' CONOPS, identify strengths and weaknesses in key logistics capabilities, and assess implications of identified strengths and deficiencies on the ability to support theater operations. It further provides logistics guidance for completion of the JSCP planning task.

It describes the LSA process, specifically within Enclosure I. The LSA provides a broad assessment of key logistics capability areas required to execute CCDRs' plans. The assessment spans the plan duration and addresses the four pillars of logistics sustainability (materiel, infrastructure, expeditionary combat support [ECS] forces, and lift). When determining requirements, a CCDR should define the total unconstrained operational logistics requirements for execution of the supported commander's CONOPS. LSA assessment should focus on key logistics support and capability areas discussed in plans. LSA findings should highlight logistics deficiencies and their associated risk to supporting theater operations. Significant deficiencies should be included in CCDRs' readiness-assessment reports and should be considered candidate issues for the joint quarterly readiness review (JQRR) action items, CCDR annual capability gap assessment, analysis in war games, and other crisis action plan assessments.

CJCSI 3141.01D, 2008

This document establishes responsibilities and procedures for the management and review of campaign and CONPLANs submitted to the CJCS. The services, USTRANSCOM, and DLA evaluate overall plan resource, logistics, mobilization, and end-to-end transportation requirements. The CCDR will prepare an LSA for each fully developed OPLAN and CONPLAN with TPFDD. The LSAs will address the sustainability for the five logistics joint capability areas (supply chain operations, operational engineering, logistics services, health service support, and operational contracting).

Detailed guidance on the preparation of the LSA will be provided by the Joint Staff J-4. The Joint Staff J-4 is responsible for providing the statement of logistics supportability, including shortfalls, impact, and mitigation strategies, to the Joint Operational War Plans Division with a copy to Director for Force Structure, Resource, and Assessment, Joint Staff.

J-4 Readiness Division (J-4 RD) serves as the primary Joint Staff J-4 point of contact (POC) for all plan reviews and coordinates with other J-4 divisions and sections to review applicable portions of plans concerning supply chain operations, operational engineering, logistics services, health service support, and operational contracting. The J-4 Supply Division serves as the POC for sustainment, WRM, munitions, POL, and contracting. J-4 Logistics Services Division serves as the POC for base operating support, including mortuary affairs. J-4 Distribution Division serves as the POC for distribution and deployment issues. J-4 Knowledge Based Logistics Division serves as the POC for logistics systems.

Additionally, Enclosure B notes that J-4 RD

> ensures that combatant commanders, services, and DLA conduct joint logistics supportability analysis for sustainment, industrial base capacity, mobility, deployment, logistics systems, engineering, and medical readiness.[2] In addition, it ensures that DCMA

---

[2] Note that DLA is included in the LSA process.

[the Defense Contract Management Agency] provides technical feasibility assessments for contracting and contract management.

CJCSM 3150.01, 1999

This document describes the Joint Reporting Structure (JRS), which exists to satisfy the NCA's need for military information to perform its functions. Although administrative in its nature and content, the document does state that JRS participants need a central catalog of recurring reports that support command decisions regarding military operations and minimize duplicative reporting and the general need for standardization in reporting systems of the Joint Staff, COCOMs, and subordinate joint forces, services, and DoD agencies. This issue has applicability to DRRS, ESORTS, Air Expeditionary Force Reporting Tool (ART), and LSA reporting requirements.

CJCSG 3501, 2008

This guide provides educational material for DoD, joint, service, and CSA senior leaders on the Joint Training System (JTS) and the Joint Training Information Management System (JTIMS) that provides automation support for JTS.[3] It calls for linking training assessment to readiness assessment by stating that a military capability is the ability to accomplish essential tasks to standard and comprises one or more of the following elements: personnel, equipment, training, supplies, and ordnance. Commanders and their staffs will use joint training assessment data to support their readiness assessment in DRRS.

Further stated is the commander's and director's commitment to assess command ability to meet joint and agency MET list (J/AMETL) standards by assessing monthly the command's proficiency using the results of training events, real-world operations, experimental events,

---

[3] The JTS is designed to improve the readiness of joint forces—that is, improve their ability to perform assigned missions under unified command.

and security cooperation activities and report MET readiness in DRRS, and to identify and report, in DRRS, program and resource shortfalls and the impact these have on the command's or agency's ability to accomplish its joint and agency training requirements.

CJCSN 3500.01, 2008

This notice provides the annual CJCS Joint Training Guidance update to all DoD components for the planning, execution, and assessment of joint training for FYs 2009–2012.[4]

CJCSM 3500.03B, 2007

This manual provides guidance to the CCDRs when implementing CJCS policy for developing J/AMETLs, planning and conducting joint training, and assessing command readiness with regard to joint capabilities. The COCOMs, services, and CSAs will use this manual when using the JTS. It provides, in great detail, a description of the JTS and command and staff responsibilities. CJCSG 3501 contains the essential information contained in this publication.

CJCSM 3500.04B, 1999

The UJTL Version 4.0 serves as a common language and common reference system for JFCs, CSAs, operational planners, combat developers, and trainers to communicate mission requirements. It is the basic language for development of a JMETL or AMETL that identifies required capabilities for mission success. At 637 pages, this document

---

[4] The objective of joint training is to improve joint readiness for future operations. Joint training programs should be developed to support mission capability requirements described in organizational J/AMETL, executed in accordance with joint doctrine, and assessed through training assessments in JTIMS to provide timely input to readiness reporting in the DRRS.

contains the tasks, measures, and criteria at the strategic national and theater levels, as well as operational and joint and interoperability tactical tasks. It also contains conditions for joint tasks by physical, military, and civil environments.

AFDD 1-1, 2006

This document provides guidance for Air Force leaders in fulfilling assigned missions. It ensures that leaders at every echelon throughout the Air Force have a baseline for preparing themselves and their forces to conduct operations. Doctrine describes the proper use of air and space forces in military operations and serves as a guide for the exercise of professional judgment rather than a set of inflexible rules. It describes the Air Force's understanding of the best way to do the job to accomplish national objectives.

The document further states the need for airmen, military and civilian, who possess the right occupational skill sets and enduring leadership competencies to form the core of force development and is the basis for all force-development efforts.[5] The construct starts with understanding mission requirements and translating them into capabilities. Doctrine takes those requirements and translates them into best practices for the service. It establishes the bedrock capabilities of the Air Force that it brings to all joint operations, within which force development is then used to create leaders and commanders. Doctrine guides the presentation and employment of Air Force capabilities.

AFDD 2, 2007

This document establishes doctrinal guidance for organizing, planning, and employing air and space forces at the operational level of conflict across the full range of military operations. It is the capstone

---

[5]   The goal of force development is to prepare airmen to successfully lead and act in the midst of rapidly evolving environments while meeting their personal and professional expectations.

of Air Force operational-level doctrine publications. These publications collectively form the basis from which commanders plan and execute their assigned air and space missions and their actions as a component of a joint service or multinational force.

One of the cornerstones of Air Force doctrine is that "the Air Force prefers—and in fact, plans and trains—to employ through a COMAFFOR who is also dual-hatted as a joint force air and space component commander (JFACC)" (AFDD 1). To simplify the use of nomenclature, Air Force doctrine documents assume that the COMAFFOR is dual-hatted as the JFACC unless specifically stated otherwise.

COMAFFOR responsibilities include organizing, training, equipping, and sustaining assigned and attached Air Force forces for in-theater missions and to maintain reachback to the Air Force component rear and supporting Air Force units; delineating responsibilities between forward and rear staff elements; and providing logistics and mission support functions normal to command.

The air and space expeditionary task force (AETF) is the organizational structure for deployed Air Force forces. Regardless of the size of the Air Force element, it will be organized along the lines of an AETF. The AETF presents a JFC with a task-organized, integrated package with the appropriate balance of force, sustainment, control, and force protection. The AETF presents a scalable, tailorable organization with three elements: a single commander, embodied in the COMAFFOR; appropriate command and control mechanisms; and tailored and fully supported forces.

The AETF will be tailored to the mission; this includes not only forces but also the ability to command and control those forces for the missions assigned. The AETF should draw first from in-theater resources, if available. If augmentation is needed, or if in-theater forces are not available, the AETF will draw as needed from the AEF currently on rotation. These forces, whether in-theater or deployed from out of theater, should be fully supported with the requisite maintenance, logistical support, health services, and administrative elements.

An AETF needs a command entity responsible for the deployment and sustainment of Air Force forces. The AFFOR staff is the mechanism through which the COMAFFOR exercises his or her service responsibilities. These sustainment activities are sometimes referred to as "beds, beans, and bullets." The AFFOR staff is also responsible for the long-range planning and theater engagement operations that fall outside the AOC's current operational focus.

The Air Force AOC provides operational-level command and control of Air Force forces and is the focal point for planning, executing, and assessing air and space operations. Although the Air Force provides the core manpower capability for the Air Force AOC, other service component commands contributing air and space forces, as well as any multinational partners, may provide personnel in accordance with the magnitude of their force contribution. The Air Force AOC can perform a wide range of functions that can be tailored and scaled to a specific or changing mission and to the associated task force the COMAFFOR presents to the JFC.

AFDD 2-4, 2005

This document is the keystone document addressing the full spectrum of ACS functions that operate in peace and in war. It stresses the need for tailored combat support packages with the airmen, facilities, equipment, and supplies required for supporting Air Force forces. It includes the following definitions:

> Agile combat support (ACS) includes actions taken to create, effectively deploy, and sustain US military power anywhere— at our initiative, speed, tempo. ACS is technologically superior, robust, flexible, and fully integrated with operations. ACS capabilities include provisioning for and protection of air and space personnel, assets, and capabilities throughout the full range of military operations.

> Expeditionary combat support (ECS) is a subset of agile combat support that responds quickly and is highly mobile. ECS is the deployed ACS capability to provide persistent and effective support for the applications of Air and Space power on a global basis.

ACS master processes apply the capability to produce the desired effects necessary to create, operate, and sustain globally responsive air and space forces.

ACS capabilities are aggregations of many activities; imbedded and cross-functional tasks performed by the 23 combat support functional areas. Collectively, the combat support functional areas generate combat capability by creating, posturing, bedding down, protecting, servicing, maintaining, and sustaining support and operational forces.

ACS is heavily dependent on integration; 23 combat support functional areas make vital contributions to Air Force operational mission capability, relying on total force (active duty, Air Reserve component, civilians, and contractors). ACS forces are organized, trained and equipped into one seamless team to optimize readiness capability and total force utilization.

Networked, adaptive combat support command and control facilitates integration with warfighting functions to optimize the commanders' ability to execute their military operation.

The defining attributes of ACS are agility, reliability, integration, and responsiveness. Agility is the attribute of ensuring timely deployment concentration, adaptive employment, and resourceful sustainment of air and space power. Reliability results from the ECS team's effectiveness, competency and health of personnel, equipment dependability, trustworthiness of information, and the consistency of ACS effects. Integration brings together or incorporates diverse parts into a common team. This is not just a combination of parts; integration creates a synergistic effect, whereby the sum is much greater than its constituent parts. Responsiveness results when critical ACS capabilities are right sized, when and where needed.

The master processes measure and answer the operationally imperative questions, such as whether the forces are ready and whether the battlespace is prepared. ACS is a key enabler in readying and preparing Air Force forces for quick response, as well as sustaining all operational activity with the right resource, at the right place, at the right time, and for the right length of time. It includes the procurement, protection, maintenance, distribution, and replacement of personnel, materiel, and installations to ensure responsive AETF support for right-sized forces supporting contingency operations. The six ACS master processes depict critical ACS capabilities that produce military readiness and responsiveness across the full range of operations:

- readying the force: ensuring force fitness and organizing, training, and equipping to provide military capability
- preparing the battlespace: assessing, planning, and posturing for rapid employment
- positioning the force: tailoring, preparing for movement, deploying, receiving, and integrating forces
- employing the force: generating timely launch or strike capability, providing right-sized essential support, and ensuring safe recovery of engaged forces
- sustaining the force: maintaining effective levels of support for global operations worldwide beginning on day one of employment operations
- recovering the force: redeployment and reconstitution, ensuring that the instrument of air and space power can effectively and repeatedly be applied at the direction of the President or SecDef.

ACS unifies the depth of support managed at all echelons of command, as well as the breadth of organic, commercial, wholesale, retail, interservice, and international environments. ACS also integrates the diverse functional areas that provide unique contributions essential to Air Force operational success, thus allowing ACS to maximize effects and making the sum more dynamic than any of its individual parts. Air Force combat support capabilities are fundamental to the success of employing air and space power.

ACS planning, execution, monitoring, and control supports the mission and provides operational risk mitigation, near–real-time combat support information, and cross-AOR resource arbitration. The key to operational risk mitigation is the integration of sustaining base and ECS capabilities for global, short-term inventory optimization. Additionally, near–real-time dynamic, continuous management of combat support information and operational intelligence ensures adaptive operations and combat support plans.

Command and control is the means by which Air Force commanders monitor and maintain situational awareness, achieve common understanding of the battlespace, assess status, plan potential courses of action, and synchronize appropriate activities to achieve effects essential to meeting military objectives. Effective command and control requires well-defined process, streamlined organization, and collaborative decisionmaking constructs that are adaptable to meet unexpected challenges. Command and control of combat support enables the commander to employ capabilities and resources effectively (despite competing demands) and provides the means for implementing combat support plans, and the agility to modify those plans as necessary to meet evolving operational requirements.

Command and control is critical to the successful employment of air and space power and should be interoperable, horizontally integrated across functions, and vertically integrated across all echelons of command, and provide organizational connectivity between commanders and decisionmakers down to the employing units. Command and control supports centralized control and decentralized execution of all combat support activities.

The COMAFFOR requires the ability to maintain awareness of the status of the blue order of battle, recognize what support capability is needed where, and direct resources accordingly. Because many Air Force resources are limited and designed to serve the needs of multiple missions in widely dispersed unified commands, centralized control and decentralized execution are especially critical to ensure an optimum balance between flexibility and responsiveness of Air Force combat support. Key to this is the concept that various echelons need visibility into and authority over assets relevant to their respective roles and responsibilities.

Full-range planning and execution of Air Force forces requires an ACS communications and information system architecture that is integrated across the functional areas and provides nonsecure and secure capability. For example, the foundation for reachback consists of command and control information centers and their supporting databases. Connectivity to any deployed operating location, including bare bases, is needed early; robust secure communications and information capabilities should connect all combat support functions. Support to achieve interoperability requires standards, frequency management, standardized systems and databases, and common processes.

The combat support center (CSC) is the strategic-level ACS node at the Pentagon. It provides global views of Air Force combat support capabilities and monitors and assesses global resource allocation by integrating multitheater requirements. The CSC also conducts integrated assessments and recommends allocation of actions for critical resources. It is the ACS component of the Air Force Operations Group (AFOG) and supports the AFOG in its mission to support the CSAF, the Secretary of the Air Force (SECAF), and the CJCS.

ACS CONOPS, 2007

A key document, it spells out the ACS concept of servicewide support. The ACS CONOPS is one of seven Air Force–level CONOPS and is the foundational combat support CONOPS of the Air Force. Air Force–level CONOPS are source documents that express a vision for how the Air Force intends to plan, prepare, deploy, employ, sustain, or recover a joint force against potential adversaries within a specified set of conditions. The ACS CONOPS provides guidance for all combat support activities in the Air Force. The ACS CONOPS timeline addresses combat support capabilities required now and as the United States moves toward 2025. It supports an AEF structure that begins before operational planning and continues through execution and sustainment of persistent operations.

The ACS CONOPS lays out the construct that describes how processes use capabilities to create support effects required for successful operational activity. It conveys how ACS master effects, master processes, and master capabilities enable the Air Force to support both in-garrison and deployed operational capabilities.[6]

The ACS goals are to make the Air Force lighter and leaner, develop a more responsive planning and execution capability, achieve an agile and effective sustainment process, and develop responsive, well-integrated ACS command and control. These are key to integrating combat support capabilities to ensure timely and persistent support for air, space, and cyberspace power to accomplish Air Force objectives. They also codify ACS best practices and lessons learned and support the Air Force Capabilities Review and Risk Assessment process. Lastly, they provide a valid, reliable, capability-based foundation for force structure decisions within the Air Force. ACS goals also provide direction for the AEF and illustrate how the Air Force community supports the national strategy delineated in the Strategic Planning Guidance (SPG), the Quadrennial Defense Review, and NMS, as well as how the Air Force deploys forces in support of the CCDRs. These goals are integral to guiding ACS to enable the operational Air Force CONOPS.

Note that there are a total of six operational CONOPS. ACS provides the support foundation and integration within and between the six operational Air Force CONOPS: global strike, global persistent attack, nuclear response, homeland defense and civil support, global mobility, and space and command, control, communications, computers, intelligence, surveillance, and reconnaissance.

ACS goals are supported with actionable, measurable objectives outlined in the ACS Supporting CONOPS (November 2007). The objectives act as major areas of concentration viewed through the lens of the doctrine, organization, training, materiel, leadership and education, personnel, and facilities (DOTMLPF) construct and clarify

---

[6] Ultimate achievement of ACS master effects is based on some assumptions. One of the most salient to the ACS project is organizational integration. Disparate functional processes must be integrated into cross-functional, process-based solutions to deliver improved critical and enabling capabilities and effects. Each component headquarters will be fully networked with its sister components' planning and execution centers, facilitating an assured, integrated common operating picture ensuring operational environment awareness and availability of decision-quality information. Such collaborative planning across all components, using robust shared visualization tools, will combine joint force capabilities while enabling real-time changes.

the ACS requirements. Each objective is traceable to at least one of the four ACS goals. The ACS CONOPS is responsible for driving capabilities-based planning (CBP) requirements toward a comprehensive roadmap for ACS capabilities. The ACS capability plan serves as that roadmap and details the direction of strategic, operational, and tactical planning and programming efforts through ACS CBP activities.

As described in the ACS CONOPS, the fundamental principles of ACS describe the basic and essential qualities and intrinsic characteristics of the integrated support provided for Air Force operational activity in both peace and war. These principles are

- properly prepared forces
- assured response
- effective beddown and sustainment
- efficient installation support
- leveraged information technology
- dependable reachback
- time-definite delivery.

Of particular interest for this analysis is the principle of reachback. Through reachback, deployed units obtain support from theater-, rear-, or CONUS-based organizations. Reachback includes requests for supplies and equipment, as well as specialty consultation and decision support to enable mission accomplishment. Deployed units transmit requests for support and status reports back through echelon chains. The status reports provide the mechanism for prioritization of requests and orders of replenishment. This process must be supported by information systems that ensure that top-priority requirements are automatically identified and delivered by the optimal transportation mode. When CCDRs require a force specialty or item, the system will reachback to CONUS and deliver the member or item where and when needed. This reachback approach will make it possible to deploy fewer functions and personnel forward for the deployment and sustainment processes. The success of reachback depends on seamless data flow from the forward location through the entire support pipeline and a correspondingly responsive pipeline flowing requested support forward.[7]

Information required supporting logistical reachback responsibilities falls into two categories: (1) anticipated service or weapon system–specific products, applications, and services and (2) unanticipated spillover tasking from the theater AOC. Weapon-specific and general ACS materiel and housekeeping supplies will be needed during any conflict or crisis employing air, space, and cyberspace power. Inventory control points (ICPs), centralized intermediate repair facilities (CIRFs), and repair depots coordinate with each theater or AOR during the peacetime/readiness phase to define specific anticipated Air Force reachback support products, services, and applications. Following COMAFFOR/A4 empowerment, the CSC/LRC requests support directly from the ICPs and repair depots. During any crisis or contingency, the theater might require support that was not anticipated previously. In such cases, the JTF/J4 can request the needed support. Requests from multiple regions that exceed available national resources will be reviewed for potential allocation decisions by the joint staff.

ACS CFMP, 2010

The commander of AFMC is responsible for implementing SECAF and CSAF decisions affecting ACS, including planning, execution, oversight, and reporting on the performance of ACS. The document that will inform and drive senior leaders' decisions is the ACS CFMP.

---

[7] The logistics support center (LSC) is another example of reachback capturing how combat support personnel use supply chain channels to acquire the necessary equipment and products needed to accomplish their assigned mission. The LSC provides deployed forces with standard base-level supply system access through a remote processing station. The system has been used extensively in exercise and contingency deployments. It mirrors a normal supply account that handles mission capable assets, replenishes readiness spares packages, provides stock control and equipment transactions, coordinates operations and maintenance and stock fund management, supports local purchase requirements, and provides a remote processing station. Other responsibilities of the LSC are to provide online computer support, centralized database access for units deploying to bare bases or collocated operating bases, and support to main operating bases as requested by the theater CSC.

AFMC is the principal proponent and subject matter expert for the development and management of the ACS CFMP.[8]

The ACS CFMP guides the development and integration of future concepts and Air Force capabilities for conducting ACS activities. It integrates ACS priorities, goals, and objectives; influences the Air Force strategic planning and programming processes; and provides a defined mechanism for better understanding of the important role of ACS across the range of military operations. The CFMP is the source document to inform Air Force leadership and guide assessments of potential integration requirements and opportunities. The ACS CFMP places primary emphasis on the capabilities that air components and their senior staffs will provide the CCDRs in support of the CCDR objectives. The intent is to link ACS goals and objectives in the Air Force corporate structure and help future commanders and leaders better understand their contributions to ACS planning, programming, and implementation efforts.

The ACS CFMP also discusses an ACS architecture. Without a fully developed architectural ACS enterprise, including the touch points to how the Air Force fights wars, requirements for ACS will continue to be incomplete, disjointed, stovepiped, and inefficient. To mitigate this risk, the ACS community—defined as including both the 28 ACS functional capabilities and the ACS organizational levels (strategic [global], operational [C-NAF], and tactical [wing])—requires architectures that document ACS capabilities, organizations, processes, information flows, technology, and the operational scenarios that use them. As a strategic resource, architectures are congressionally mandated to support federal, DoD, joint, and Air Force requirements, as well as interoperability.

ACS architectures, as a part of the federated Air Force Enterprise Architecture, are a key framework component that connects ACS capabilities to the Air Force's strategic vision, strategies, and plans. The ACS

---

[8] ACS is one of the 12 Air Force service core functions (SCFs). Underpinning the work of all Air Force SCFs are the capabilities included in ACS. The SECAF and the CSAF designated the role of Air Force SCF lead integrators to specific MAJCOMs to lead development and advancement of Air Force positions on the SCFs. AFMC is assigned the lead integrator role for ACS.

architecture should also capture potential gaps, shortfalls, and system duplication, as well as support functionality and system interoperability investigations through architecture-based analysis. This architecture-derived information is critical to supporting high-level planning, programming, and requirements decisionmaking. A fully documented (architected) ACS community will describe the community's current operational capability and be used to identify metrics for implementing near- and mid-term modernization efforts.

AFDD 2-8 (now AFDD 6-0), 2007

This doctrine document is a keystone doctrine statement and establishes doctrinal guidance for command and control operations to support national military objectives and commanders in employing air and space forces across the full range of military operations. It stresses the need for fixed and mobile interoperable command and control centers, with efficient processes, state-of-the-art equipment, and properly trained airmen to support U.S. and multinational requirements worldwide.

Paramount to Air Force command and control doctrine and stated in AFDD 2-8 is the fact that effective command and control is essential to the Air Force in producing the right effects at the right place and time to support theater and global force commanders.

According to AFDD 2-8, command and control involves the integrated processes, organizational structures, personnel, equipment, facilities, information, and communications designed to enable a commander to exercise authority and direction across the range of military operations. The command and control process should support informed and timely decisions at all levels of command.

Specifically addressed are ACS planning, execution, monitoring, and control: Combat support command and control enables the commander to employ capabilities and resources effectively (despite competing demands). It also provides the means for implementing combat support plans and the agility to modify those plans as necessary to meet evolving operational requirements.

ACS planning, execution, monitoring, and control processes and capabilities are described. ACS uses the MAPE process. ACS systems provide the tools and technology to access, analyze, display, and act on relevant information enabling them to ready, deploy, employ, and sustain forces for assigned missions worldwide.

Also discussed in AFDD 2-8 is that Air Force command and control centers must provide oversight and control for operations conducted worldwide. These centers must support both the operational and administrative chains of command. Each command and control center is unique in its mission due to the uniqueness of the command and the mission that it serves. Air Force command and control centers may use the concepts of reachback and distributed operations to support forces deployed or operating in place from multiple locations.[9]

Reachback provides ongoing combat support to the operation from the rear, while *distributed operation* indicates actual involvement in operational planning or operational decisionmaking. The goal of effective distributed operations is to support the operational commander in the field; it is not a method of command from the rear. The concept of reachback allows functions to be supported by a staff at home station, to keep the manning and equipment footprint smaller at a forward location. Distributed operations, which may rely heavily on reachback support, vary by mission, circumstances, and level of conflict. Each Air Force command and control entity will have a defined function that contributes to an overall distributed operation, whether it provides information from a fixed location at home station or it is forward deployed.

---

[9] *Reachback* is a generic term for obtaining forces, materiel, or information support from Air Force organizations not forward deployed. The intent of reachback operations is to support forces forward, not to command operations from the rear. Distributed operations occur when independent or interdependent nodes or locations participate in the operational planning or operational decisionmaking process to accomplish goals or missions for engaged commanders.

Command and Control Service CFMP, 2010

This plan provides the Air Force's path required to achieve its command and control vision. It identifies how the Air Force will deliver command and control across the full spectrum of conflict, implications and risks generated by current and future operational environments, and DOTMLPF actions designed to mitigate noted implications and risks to ensure that the Air Force can deliver command and control to the joint force. The command and control SCF plan broadly describes strategic-level guidance for Air Force leadership to provide command and control capabilities to the JFC for application across the range of military operations.

Commander, ACC (COMACC), is the lead integrator for the command and control CFMP and oversees foundational analysis, development, and refinement. The challenge in defining the command and control SCF is to recognize and acknowledge that Air Force command and control occurs across all MAJCOMs, C-NAFs, and wings and is executed by individual weapon system operators. It occurs in all warfighting domains and in all stages of military operations.

PAD 10-02, 2009

This document discusses command and control of Air Force forces at the operational level through the C-NAFs, composed of an AOC and an AFFOR staff. It scopes the C-NAFs' day-to-day responsibilities for phase 0 (shape) and phase 1 (deter) operations and establishes the requirement for rapid augmentation teams to meet surge requirements and supplement C-NAFs during other phases of operations. This PAD further defines the number and location of the AOCs, including their required architecture and training suites, and aligns reserve units with C-NAFs for augmentation. This document also directs the development and testing of an OSF. The OSF was established (per PAD 06-09) to provide COOP and backup to C-NAFs conducting ongoing operations. It was also designed to provide reachback support to both the

AOC and AFFOR staff. This PAD directs testing the OSF concept and developing TTP for distributed operations.

AFPD 10-2, 2006

This policy directive establishes Air Force readiness requirements and responsibilities and directs MAJCOMs to report accurate readiness data in support of decisionmaking processes.

AFPD 10-2 states that all Air Force readiness-related programs and processes will be aligned with DRRS (GFM) initiatives. DRRS will feed ESORTS data to facilitate the JQRR process. GSORTS is a CJCS-controlled, automated data system primarily created to provide the President and the SecDef authoritative information related to the readiness of military forces to meet assigned missions and goals.

The ART allows AEF allocated units the ability to report UTC-level readiness data. It allows immediate updates and ready access to an aggregate UTC status for all levels of command with sufficient depth of information to make informed decisions on the employment of Air Force capabilities across the full ROMO. Integration of DRRS and ART (DRRS-AF) is critical to provide the required visibility of Air Force capabilities and resources while supporting the AEF construct.

APPG, 2009

This biennial document provides both top-level planning direction to MAJCOMs and programmatic guidance to the Air Force corporate structure (AFCS). This APPG is oriented toward developing planning activities that will enable the Air Force to construct a well-informed, strategy-driven, resource-constrained FY 2012 POM. The APPG is the primary document directing disciplined planning efforts across the Air Force.

DRRS Interim Guidance, 2009

This guidance, issued by AF/A3O to implement DRRS reporting by Air Force units, is issued without applicable DoD instruction (DODI) and CJCSI. When the DODI and CJCSI are published, this guidance will be published in a corresponding AFI. Until that occurs, this guidance will be used. If there is a conflict between this guidance and the CONOPS for DRRS ESORTS, this guidance takes precedence until all documents are updated.

AFUTL, 2009

The AFUTL is a menu of tasks in a common joint language that serves as the foundation for CBP across the range of military operations. It is a comprehensive, integrated menu of functional tasks, conditions, and measures supporting all levels of DoD in executing the national defense strategy and the NMS. The AFUTL supports DoD in joint CBP, joint force development, and readiness reporting.

Air Force–specific tasks supplement the UJTL and, when they are combined with the appropriate tasks in the UJTL, the two comprise the AFUTL. This task list also complies with DODD 7730.65, *Department of Defense Readiness Reporting System (DRRS)*, for developing and assessing mission essential task lists (METLs) in DRRS.

The AFUTL is a key element of the capability-based, mission-to-task joint system. The AFUTL is adaptive, flexible, and horizontally and vertically integrated, and tasks are mapped to capabilities to meet operational mission requirements. This capability-based, mission-to-task connectivity enables combat developers to determine what DOTMLPF changes affect future force development. Additionally, these tasks enable operational planners to determine what forces are required to achieve desired capabilities when used in conjunction with DRRS. All users of this list must first conduct unit mission analysis, identify specified and implied tasks, then use the AFUTL or UJTL to describe these tasks (including supporting and command-linked tasks). Users then apply guidance to determine which tasks are essential to successful accomplishment of that mission ("war stopper"). These are a unit's METs. Mission tasks are further described by selecting the conditions under which the task is to be performed and the measures and scale of task performance, forming a unit METL.

AFPD 10-4, 2009

This AFPD directs implementation of U.S. Code Title 10 guidance on service responsibilities. It covers mostly personnel and manpower utilization.

This policy directive incorporates aspects of global operations planning and directs AEF force management, force allocation, and battle-rhythm management synchronized with the GFM process to support CCDR requirements. This AFPD aligns actual tempo requirements with a matching force-generation model under a tempo band construct, applies rule sets for forces operating at a tempo of less than 1:4 deploy-to-dwell, aligns Air Force planning periods with GFM, provides an enterprise view of service risk synchronized with GFM risk definitions, and directs the annual reassessment of planning assumptions based on emerging and rotational requirements.

AFI 10-201, 2006

The document addresses Air Staff FAM responsibilities: Develop measured area criteria and their associated tables and conversion charts, as applicable; ensure that tables and conversion charts are current and accurately reflect the functional area's mission; monitor functional-area Status of Resources and Training System (SORTS) information; coordinate changes affecting SORTS; ensure that equipment and supplies identified in supported designed operational capability (DOC)–referenced UTCs are not selectively excluded from SORTS measurement; provide guidance to MAJCOM FAMs for construction and maintenance of unit capability (that is, UTCs) and the associated SORTS DOC statement; coordinate among UTC providers to ensure standardization of capability; monitor and anticipate changes in capability or wartime requirements and direct necessary modifications; resolve MAJCOM disputes concerning new UTC or special mission capability additions to the SORTS DOC statement; coordinate any required interim SORTS guidance; identify IOC dates for new SORTS tasks; and periodically (minimum, semiannually) HAF functional offices for accuracy.[10]

AFI 10-401, 2006

The purpose of this instruction is to provide an overview of the joint planning process and the interrelationships of the associated national-level systems that produce national security policy, military strategy, force and sustainment requirements, and plans.[11] The four major interrelated systems affecting the development of joint operational plans are the National Security Council System; the JSPS; Planning, Pro-

---

[10] SORTS is an internal management tool for use by the CJCS, services, unified commands, and CSAs. It is the single automated reporting system within DoD functioning as the central registry of all operational units of the U.S. armed forces and certain foreign organizations. SORTS has a threefold purpose: It provides data critical to crisis planning, provides for the deliberate or peacetime planning process, and is used by CSAF and subordinate commanders in assessing their effectiveness in meeting Title 10 responsibilities to organize, train, and equip forces for COCOMs. The Air Force uses SORTS status information in assessing readiness, determining budgetary allocation and management action impacts on unit-level readiness, answering congressional inquiries, analyzing readiness trends, and supporting readiness decisions. SORTS is not designed to function as a detailed information management system objectively counting all conceivable variables regarding personnel, training, and logistics. SORTS indicates a unit's ability to undertake its full mission (primary DOC statement) or particular mission (secondary or tertiary DOC statements). SORTS may also provide indications of the efficacy of resource-allocation decisions and the impacts of budgetary constraints on resourcing unit requirements. When deployed or employed in response to a crisis or OPORD, SORTS provides both an assessment of a unit's status based on ability to execute the mission set for which it was organized or designed and, when appropriate, the mission against which it is employed.

[11] The Joint Programming Guidance is discussed. It provides programming guidance to military and defense agencies to develop their POM. It also provides the SecDef's threat assessment, policy, strategy, force planning, and resource planning guidance within broad fiscal constraints. It is the link between the JSPS and the PPBE.

gramming, Budgeting, and Execution (PPBE); and the JOPES. This instruction provides very detailed planning guidance.

The PPBE is a major system related to the overall joint planning and execution process. Planning, programming, and budgeting are ongoing processes that enable senior leadership to assess alternative ways to achieve the best mix of force, requirements, and support attainable within fiscal constraints. A major goal is to strategically link any major decision for acquisition, force structure, operational concepts, and infrastructure, for example, both to the JPG and to program and budget development. The PPBE is concerned with allocating resources (force, equipment, and support) to meet the warfighting needs of the CCDRs. It translates strategy and force requirements developed by the military in the NMS into budgetary requirements that are presented to Congress. Key products in the PPBE include the POM, Budget Estimate Submission, the President's Budget, Program Change Proposal, and Budget Change Proposal.

Also described in this AFI is GFM as a process to align force apportionment, assignment, and allocation methodologies in support of the defense strategy and in support of joint force availability requirements; present comprehensive insight into the global availability of U.S. military forces; and provide senior decisionmakers a vehicle to quickly and accurately assess the impact and risk of proposed allocation, assignment, and apportionment changes.

In this process, the Joint Staff directly tasks USJFCOM, as primary JFP, or other JFPs (i.e., USTRANSCOM, USSOCOM, or U.S. Strategic Command [USSTRATCOM]), to develop recommended global sourcing solutions. This formal process sources emerging unified CCDR (unified combatant commander) requirements. USJFCOM endorses the request for forces/capabilities (RFF/RFC) and forwards with any additional sourcing guidance to the service components to determine sourcing recommendation and issues. The Air Force Operations Group (AF/A3OO), working with HAF and MAJCOM FAMs, will develop and provide the Air Force position to ACC as the USJF-COM service component to distribute the recommended sourcing solution (including RCs) to the primary JFP. ACC has a clear component role with USJFCOM to assist in distributing and monitoring

Air Force sourcing availability to execute CCDR missions and forecast sourcing challenges or issues.

RFFs/RFCs and the Joint Staff process are initiated when a CCDR determines a requirement and submits an RFF/RFC to the Joint Staff. The RFF/RFC provides CCDRs with a means to obtain required support not already assigned or allocated to the command. Prior to the CCDR forwarding to the Joint Staff, Air Force component headquarters will review all RFFs/RFCs for Air Force capabilities being requested and translate the request into potential UTCs, Air Force specialty codes, or closest organic Air Force capability.

In response to an RFF/RFC, the Joint Staff generates a draft execute or deployment order (EXORD/DEPORD) allocating forces from a force provider to the requesting CCDR for a set period of time and sends the validated RFF/RFC in a Joint Staff Action Package to USJFCOM and a copy to HAF and supporting COCOMs. Generally, this is a three-step Air Force service-level process beginning with sourcing feasibility (action officer to action officer), formal sourcing, and culminating with an EXORD/DEPORD or modification to EXORD.

The Air Force–recommended sourcing solution is delivered to ACC, as Air Force JFP, to USJFCOM for final input and to prepare rotational force schedule, rotational force allocation plan, and military risk assessment for the Joint Staff to submit to the SecDef for approval.

For Joint Staff taskings, AF/A3OO coordinates with HAF and MAJCOM FAMs and the AEFC to accurately assess availability of assets, the risk associated with sourcing the requirement, necessary training, and appropriate latest arrival date. AF/A3OO forwards a recommendation through Operational Planning Policy and Strategy, Headquarters Air Staff to ACC, which forwards a fully coordinated response to USJFCOM.

ACS sustainment planning is a crucial element of crisis action and contingency planning. The Air Force accomplishes this planning by means of an LSA. LSA is an analytical process used to predict ACS operational capability requirements, gaps, and priorities. The process and methodology support Defense Planning Guidance and major theater OPLAN assessments, crisis action planning, and supplemental budgeting estimates. Air Force Materiel Command, Logistics Readiness Division, validates all logistics planning factors developed by Air Force and other DoD organizations. AF/A4/7 reviews these planning factors to ensure that they are consistent with policy guidance, ACS CONOPS objectives, and capability review and risk assessment scenarios and priorities. This assessment provides a broad assessment of key ACS support and enabler capabilities required to execute the DPG and CCDR's plans. As a general rule, the Air Force uses the supported component headquarters' directorate of logistics, or equivalent, as its agent for analysis.

AFI 10-403, 2008

This instruction implements AFPD 10-4, *Operations Planning: Air and Space Expeditionary Force (AEF)*, and AFI 10-401, *Air Force Operations Planning and Execution*. It provides the basic requirements for Air Force deployment planning and execution at all levels of command to support contingency and deployment operations. It also describes the specific requirements for preexecution, command and control, cargo and personnel preparation, reception, support of Air Force deployment and redeployment operations, and reintegration and reconstitution procedures. This guidance directly supports the installation commander to effectively and efficiently deploy forces in support of OPLANs, AEF taskings, lesser contingency operations, exercises, and other operational and training events.

The AF/A3O is responsible for overall readiness and training of Air Force forces, contributing to a force that is trained and ready to deploy. It publishes direction concerning the ART allowing unit commanders to assess a UTC's ability to meet its mission capability and the AEFC to task the most-qualified and most-ready units.

The AF/A4/7 provides policy guidance to the Air Staff, MAJCOMs, and wings to achieve effective and efficient worldwide deployment of identified forces comprised of required capabilities and their inherent resources.

The Directorate of Logistics Readiness is the office of primary responsibility (OPR) for Air Force deployment and redeployment operation and develops policy guidance to support Air Force deployment objectives.

The AF/A4/7 does the following:

- develops training curriculum for installation deployment officers (IDOs), including cascade training for unit deployment managers (UDMs)
- develops policy guidance for integrating automated systems to support deployment operations and serves as the ACS OPR for Deliberate and Crisis Action Planning and Execution Segments (DCAPES)
- serves as the overall OPR for the Integrated Deployment System (IDS). IDS is composed of four information technology systems: the Logistics Module (LOGMOD), Cargo Movement Operations System (CMOS) or Global Air Transportation Execution System at AMC CONUS strategic aerial ports and OCONUS Air Mobility Squadrons, Automated Air Load Planning System (AALPS), and DCAPES.
- manages the LOGMOD/LOGMOD Stand-Alone, CMOS, and AALPS components of IDS
- develops policy guidance on transportation activities that support deployment operations
- develops policy guidance on automated cargo and passenger transportation systems to support deployment processing and in-transit visibility of deploying personnel and cargo
- reviews and approves or disapproves logistics detail additions, deletions, and changes and forwards those changes to the Joint Staff for update to the type unit characteristics
- serves as core member of the AEF Steering Group (AEFSG) developing policy and procedures to enhance the execution of the AEF.

The Directorate of Maintenance develops policy guidance on maintenance organizations' support of deployment operations.

The AF/A7C establishes and maintains civil engineer policy and guidance to ensure that civil engineers have the capability to provide, sustain, operate, maintain, restore, and protect the installations, infrastructure, facilities, housing, and environment necessary to support air and space forces involved in deployment, sustainment, and redeployment operations.

COMACC is responsible for execution of the AEF schedule. COMACC performs the following functions in overseeing the scheduling and execution of the AEF:

- performs as the Air Force JFP recommending global Air Force sourcing solutions to USJFCOM
- manages the scheduling and sourcing of forces to meet AEF requirements through the AEFC
- ensures that MAJCOMs verify status of UTCs in ART
- adjudicates issues that cannot be resolved by the commander, AEFC (AEFC/CC) and affected air component or MAJCOM commanders
- forwards the fully coordinated AEF schedule through the Deputy Chief of Staff, Operations, Plans, and Requirements (AF/A3/5) to CSAF for approval to meet GFM-directed timelines for inclusion in the GFM guidance
- task organizes an AETF from scheduled forces and will pass the sourcing requirement to the affected MAJCOM on receipt of an HAF tasking order
- Task organization and transfer of AETF forces are coordinated through COMACC and the AEFC.
- passes (through AEFC) sourced capability to the affected MAJCOM commander (MAJCOM/CC) for execution
- provides HAF (through AEFC) visibility over deployed forces to assess location, readiness, and projected reconstitution requirements.

AEFC is a direct reporting unit of AFPC and manages and coordinates the AEF schedule and tracks execution.

AEFC is a service organization and is itself without authority to exercise operational authority over forces. Rather, AEFC facilitates the transfer of forces. It coordinates efforts of the scheduling integrated process teams (SIPTs). Each of the affected MAJCOMs will have appropriate representation on the SIPTs and will coordinate on SIPT actions. The three SIPTs are

- Combat Air Forces SIPT (CAF SIPT), chaired by ACC A3
- Mobility Air Forces SIPT (MAF SIPT), chaired by AMC A3
- ECS SIPT, chaired by AEFC/CC.

The AEFC

- manages the ART
- provides a monthly report of UTCs in their eligibility period and not reporting "green" in ART through AF/A3/5 to CSAF
- assists the C-NAFs in identifying capabilities and UTCs required in the AETF
- maintains the master rotational TPFDD by building requirements after initial TPFDD build by the C-NAFs
- upon receipt of CSAF execution order, passes sourcing recommendations in accordance with the AEF schedule to the affected MAJCOM/CC (information copy to units) for execution
- works with COMAFFORs and USTRANSCOM to maintain in-transit and deployed visibility and tracking of AEF UTCs
- monitors the scheduling of deployment transportation
- manages the DCAPES tasking process for AEF-sourced requirements as identified in contingency and rotational TPFDDs
- manages the Air Force Deployment Processing Discrepancy Reporting Tool (DPDRT)
- articulates related processes, roles, and responsibilities of all involved agencies (i.e., the AEFC, MAJCOMs, IDOs, personnel readiness functions, UDMs, personnel support for contingency operations teams, and deployed commanders)
- maintains the DPDRT and produces metrics to track and report discrepancies for corrective actions
- is responsible for monitoring the corrective action taken by the supporting commands through the DPDRT program
- oversees and manages the UTC shortfall and reclama process when taskings must be reassigned between MAJCOMs; the

AEFC is the central agency for adjudicating Air Force reclamas and forwarding to HAF, as required
- will designate ECS backfill forces when required; these forces can be deployed, or placed on a prepare-to-deploy order, as appropriate, if theater-assigned forces are unable to disengage to respond to an unexpected crisis in their assigned theater
- serves as co-chair of the AEFSG developing policy and procedures to enhance the execution of the AEF.

AFI 10-244, 2005

This instruction implements AFPD 10-2, *Readiness*. It provides policy and guidance for reporting Air Force UTC status. It formalizes reporting policies for taskings for the full range of military operations.

This AFI discusses ART. The ART allows AEF-allocated units the ability to report UTC-level readiness data. It provides one central location to archive reported data. It allows immediate updates and ready access to an aggregate UTC status for all levels of command with sufficient depth of information to make informed decisions on the employment of forces for AEF operations. It further provides a means for identifying and analyzing actionable indicators of change.

ART complements readiness data reported in SORTS. ART focuses reporting on the modular scalable capability-based UTC's designed to meet the needs of the 21st-century force, while SORTS is unit-centric with reporting based on major war commitments.[12]

This AFI outlines roles and responsibilities as follows.

---

[12] Readiness assessments for major war and AEF tasking must be considered together; however, the reporting guidelines for each may be independent. A unit's C level as reported in SORTS might not directly correlate to its ability to support a specific UTC tasking as indicated in ART. Unit commander assessments reported in ART present the status of each UTC in the AEF library, and they provide higher levels of command the necessary information to make force and resource-allocation decisions to effectively support theater commanders. Within the AEF construct, the UTC assessments are used to determine the most effective force tasking.

- The AF/A3/5 coordinates Air Force–wide efforts to develop capabilities and field AEF forces and the associated operational-level command and control infrastructure and units.
- The AF/A5X assesses capability of AEF forces to support CCDR planning initiatives and requests for support and assesses capability of apportioned AEFs to accomplish assigned missions. It interfaces with the AFPC Directorate of AEF Operations (AFPC/DPW) on UTC efforts, AEF libraries, JSCP issues, AEF sourcing issues and conferences, and FAMs' interface with and AEFs' relationship to OPLAN guidance. It is the Air Force focal point for developing and integrating operational strategies, requirements, policies, guidance, and plans necessary to support AEF operations worldwide supporting the warfighter. The War and Mobilization Planning Policy Division develops general policies regarding all facets of the management of UTCs and the general guidelines for assigning available UTCs to the AEF construct.
- The AF/A3O is responsible for overall Air Force current operations, readiness, and training. It administers policies governing operational training, force readiness, range and airspace issues, personnel recovery, and special plans and programs. It is the OPR for Air Force readiness.
- The AF/A4/7 develops policy and provides guidance for all logistics plans, transportation, supply, maintenance, civil engineer, and munitions support. It is the HAF lead for developing ACS capabilities and appropriately sizing these capabilities as ECS to support AEF operations.
- The Air Staff FAMs act as a central coordinator of the actions of their MAJCOM, FOA, and DRU counterparts to ensure that their applicable functional-area UTCs are being properly assigned to the AEF construct. (See AFI 10-401 for additional Air Staff FAM responsibilities.)
- The AFPC/DPW is a cross-functional, centralized management team responsible for planning, configuring, scheduling, and preparing AEFs, as well as assessing AEF capabilities to enable the advancement of the AEF. Responsibilities specifically include AEF force tasking and scheduling for steady state operational require-

ments. The AFPC/DPW integrates trained aerospace forces to meet theater CCDRs' requirements. Included in this is responsibility for working with the Air Force Operations Group during crisis action planning and with AF/A3/5 for force reconstitution planning. It identifies escalated reconstitution requirements when force commitment exceeds sustainment levels. It coordinates with MAJCOMs, FOAs, and DRUs to identify units in surge operations and those that require reconstitution. It monitors personnel, training, equipment, and supply status throughout surge operations, advising Air Staff of critical impacts to on-call operations, the AEF schedule, and MCO execution.

- The AFPC/DPW monitors UTC readiness through ART. It assesses UTC problem areas for overall AEF impact. It assists in asset reprioritization based on reported UTC readiness level. It monitors UTC shortfalls and deficiencies and ensures visibility by MAJCOMs, FOAs, and DRUs and Air Staff FAMs.

AFI 13-1AOC, Vol. 3, 2005

This instruction implements guidance in JP 3-30, *Command and Control for Joint Air Operations*, and AFDD 2, *Operations and Organization*. It covers the AOC weapon system that is provided by the AFFOR staff and employed by the COMAFFOR when designated as a combined or joint force air component commander (C/JFACC), supporting component commanders or when executing air and space operations and no C/JFACC is designated.

The AOC weapon system is the senior command and control element of the Theater Air Control System and operational-level focal point for command and control during Air Force and combined (coalition) or joint operations. Following the tenet of centralized planning and control, and decentralized execution, the AOC provides operational-level command and control of air and space forces. The AOC includes personnel and equipment to ensure the effective conduct of air and space operations (for example, communications, operations, intelligence).

The combat support team (CST) is the AOC's focal point for all combat support–related issues affecting the AOC's processes and the support and sustainment of Air Force, joint, and coalition combat air power. The CST directly participates in strategy development, combat planning, combat operations, and air mobility planning and operations. The CST analyzes COAs for feasibility and assesses the logistical and combat support feasibility of each COA. During strategy development and other planning processes, the CST assesses the potential impact beddown decisions and assesses the impacts of TPFDD feasibility and force closure estimates. The CST supports all AOC sections with combat support information. The CST provides data for daily J/CFACC decision and status briefings and information for recurring reports. The CST maintains logistics status reports and combat support status reports and ensures that automated planning systems data (e.g., status of aircraft, facilities, medical, munitions, personnel, POL, and resources) and all other combat support–related decision-support tools reflect the most-current information. As a minimum, the CST is comprised of aircraft maintenance, civil engineer, logistics readiness, and munitions officer and enlisted personnel but can be expanded with representation from other COMAFFOR staff directorates as required.

Joint and Air Force Command Structure

A CCDR commands either a geographic command (such as USCENTCOM, USPACOM, or USEUCOM) or a functional command (such as USTRANSCOM or USSTRATCOM). That CCDR maintains operational, tactical, and administrative control over troops operating in his or her AOR. A representative from each service component—Army, Navy, Air Force, and Marines—reports directly to the CCDR to help achieve his or her campaign objectives. The Air Force representation to the CCDR is the COMAFFOR. During military operations, the CCDR can name a JTF commander to carry out operations in the AOR. In this case, the COMAFFOR would report to the JTF commander (see Figure C.1), who would, in turn, report to the CCDR.

The COMAFFOR plans and executes all Air Force air and space operations in the AOR. He or she is the Air Force component representative to the COCOM. The vision behind the present approach for providing a single Air Force voice to the COCOM is summarized in PADs 06-09 and 07-13 and the *Air Force Forces Command and Control Enabling Concept* (PAD 06-09, 2006b; PAD 07-13, 2008; U.S. Air Force, 2006a). These documents outline guiding principles that indicate that each command—and, where appropriate, each subunified command—will have a single POC for U.S. Air Force forces, with COMAFFOR authority. Under the current organizational structure, the COMAFFOR is either the component major command com-

Figure C.1
Typical Unified Combatant Chain of Command

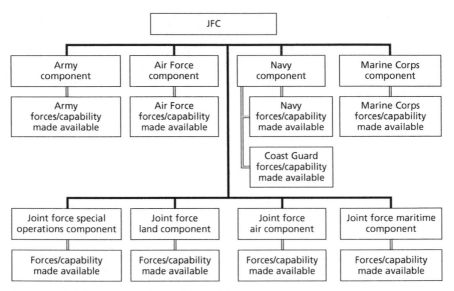

SOURCE: JP 1 (2007[2009]), Figure V-1, p. IV-3.
RAND MG1070-C.1

mander (C-MAJCOM/CC) or the C-NAF commander (C-NAF/CC)
(see Figure C.2).[1]

In addition to planning and executing all Air Force operations in
theater, the COMAFFOR has responsibility for the care and feeding
of all Air Force personnel engaged in operations in the AOR. To help
him or her fulfill these responsibilities, the COMAFFOR commands
two organizations: the AOC and an AFFOR staff (see Figure C.3). The
AOC typically concentrates on prosecuting the operation. The CCDR
sets objectives, and the AOC responds with available Air Force capabilities and builds an ATO to accomplish those objectives (operational-
level command and control). The AFFOR staff primarily concentrates
on enabling the forces to accomplish the assigned missions by ensuring
that all required support is available (care and feeding).

---

[1] C-MAJCOMs are USEUCOM, USPACOM, USSOCOM, and USTRANSCOM.
C-NAFs are USCENTCOM, U.S. Northern Command (USNORTHCOM), U.S. Southern Command (USSOUTHCOM), and USSTRATCOM.

Figure C.2
Air Force Component Headquarters Templates

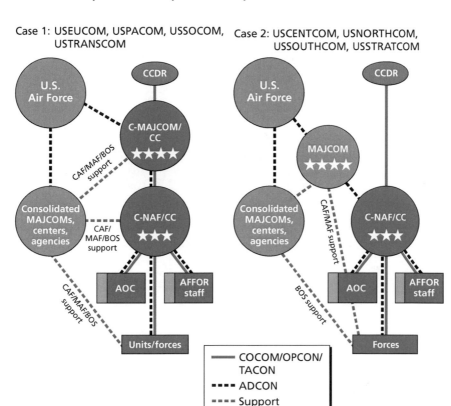

SOURCE: U.S. Air Force (2006a), Figures 2 and 3, p. 10.
RAND *MG1070-C.2*

The COMAFFOR focuses primarily on meeting the needs of the JTF commander. He or she should be postured to assume JFACC duties as needed, which would include command of the AOC as well as a joint staff working in parallel with the existing AFFOR staff. The AOC should be capable of quickly incorporating joint elements necessary to perform JFACC tasks. The joint staff should be postured to answer joint taskings, while the AFFOR staff remains an Air Force headquarters staff working for a COMAFFOR (either the C-NAF/CC or a C-MAJCOM/CC) even if the COMAFFOR assumes JFACC authority.

Figure C.3
Component Numbered Air Force Organizational Construct: Air and Space Operations Center and Air Force Forces Staff

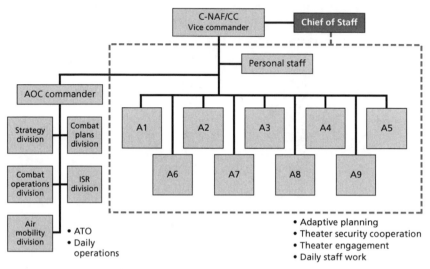

SOURCE: U.S. Air Force (2006a), Figure 5, p. 17.
NOTE: A2 = intelligence directorate. A3 = air, space, and information operations directorate. A8 = strategic plans and programs directorate. A9 = analyses, assessments, and lessons learned directorate.
RAND *MG1070-C.3*

The AFFOR component of each C-NAF is comprised of the AFFOR staff and a personal staff. The AFFOR staff has primary responsibility for shaping, posturing, and sustaining Air Force forces for employment. The AFFOR staff supports the COMAFFOR in planning and executing component and operational tasks, while the personal staff helps with matters requiring close personal attention by the commander.[2] The personal staff also advises the commander on technical and administrative matters. AFFOR staff tasks include short-term responses to the immediate needs in crisis or steady state operations and longer-term tasks associated with creating, posturing, and sustaining a combat force. The AFFOR staff is the Air Force component-level staff.

---

[2] See AFDD 2 (2007a) for a description of AFFOR functional responsibilities.

There are currently 11 C-NAFs located around the world. Figure C.4 shows the seven geographic C-NAFs—Air Forces Northern (AFNORTH), AFCENT, AFEUR, AFAFRICA, AFKOR, AFPAC, and AFSOUTH—and the four functional C-NAFs—AFSOC, Air Forces Transportation (AFTRANS), Air Forces Strategic–Space (AFSTRAT-SP), Air Forces Strategic–Global Strike (AFSTRAT-GS)—as well as other NAFs.

All these organizations and personnel—the COCOM and CCDR, the JTF and JTF commander, the COMAFFOR, the C-NAFs, and the C-MAJCOMs—play an important role in ACS planning, execution, monitoring, and control processes. However, the ACS organizational structure used in the past several military operations were developed ad hoc, contributing to several of the organizational deficiencies listed in Chapter Six.

Doctrine calls for a NAF to transition to the wartime Air Force component role in times of conflict. Doctrine also calls for the augmentation of the NAF for reachback capability. During JTF Noble Anvil (JTF NA), the air war over Serbia, the Air Force deviated from doctrinal guidelines and placed the AFFOR and JFACC staffs at separate locations. The 16th Air Force commander was selected to be the JFACC. The 16th Air Force A4 was quickly overwhelmed by his responsibilities and looked to the MAJCOM component, U.S. Air Forces Europe (USAFE), to provide support. At the beginning of JTF NA, USAFE did not have a clearly established role to execute contingency responsibilities. The staff faced challenges in organizing to provide this support. Their data gathering and analysis had varying degrees of success with respect to data accuracy and timeliness. They struggled to estimate support needs and present them to USEUCOM. As JTF NA progressed, ACS planning, execution, monitoring, and control organizational roles and responsibilities evolved.

As in JTF NA, combat support command relationships during OEF did not follow doctrine. Doctrine called for augmenting AFCENT logistics personnel; elements of the AFCENT logistics directorate deploying forward, if forward operations were necessary; and a reachback logistics element at the AFCENT rear site at Shaw AFB in South Carolina. Instead of augmenting the NAF, the AFCENT logistics and

Figure C.4
Air Force Component Headquarters Construct

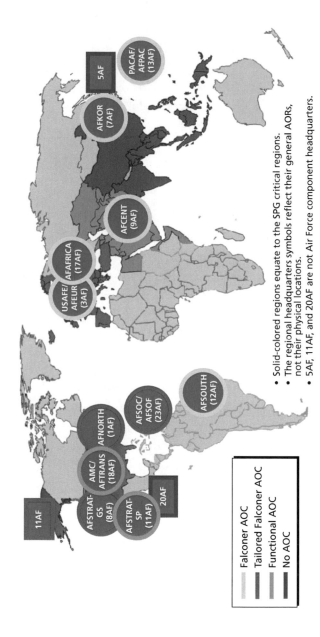

- Solid-colored regions equate to the SPG critical regions.
- The regional headquarters symbols reflect their general AORs, not their physical locations.
- 5AF, 11AF, and 20AF are not Air Force component headquarters.

Legend:
- Falconer AOC
- Tailored Falconer AOC
- Functional AOC
- No AOC

SOURCE: U.S. Air Force (2006a), Figure 1, p. 9.
NOTE: AFSOF = Air Force special operations forces. PACAF = Pacific Air Forces.

RAND MG1070-C.4

installation personnel, with the ACC logisticians and civil engineers, established augmentation arrangements with ACC at Langley AFB in Virginia. The logistics and installation rear element reached back to the ACC Logistics Readiness Center and Contingency Response Cell for needed staffing. Langley supported the logistics and installation reachback responsibilities of the parallel AFCENT directorates, which went forward to Prince Sultan Air Base in Saudi Arabia to work with the COMAFFOR/JFACC.

The augmentation arrangement was pulled together at the last minute, creating several challenges. First, the augmentation personnel were not familiar with theater-specific plans, limiting their effectiveness in carrying out responsibilities that relied on in-depth knowledge of threat, host nation, and theater issues. Second, augmentation personnel were not always familiar with command-unique policies and procedures (many of which are undocumented and have evolved from personal relationships between staff members and intratheater agencies), and they lacked training on locally developed decision-support tools that are prevalent in the absence of standardized information systems. Finally, augmentation personnel might lack experience with the core staff they are joining and hence might not contact the most knowledgeable person when seeking help.

At the beginning of OEF, the AFCENT communications and information directorate already had forward and rear elements in place for Operation SOUTHERN WATCH. These elements were augmented during OEF by activating parts of the ANG associated with AFCENT and with ACC Director of Communications and Information. The communications and information directorate reachback originally operated from Shaw AFB. A few weeks into the operation, staff relocated to Langley AFB to foster closer integration with the ACC Crisis Action Team Support Cell but later moved back to Shaw AFB.

In addition, the AFFOR logistics and installation functions were performed in the CAOC, although not corresponding to their doctrinal responsibilities. Doctrine calls for the separation of the AFFOR staff and CAOC functions. Although having AFFOR logistics and installation functions in the CAOC was not technically a break in doctrine, division of duties and responsibilities could be confused because of this collocation. Traditionally and according to doctrine, the CAOC is responsible for developing the ATO. Consequently, the logistics contingent in the CAOC was responsible for assessing resources needed to support the ATO and the effects of any resource shortages. The AFFOR logistics staff, on the other hand, needed to concentrate on assessing support effectiveness of alternative deployment and employment concepts identifying constraints to the operations and plans staff and the AOC, although they do not work for the AOC/CAOC.

As learned in JTF NA and OEF, ACS processes were not well documented in either Air Force or joint doctrine. Doctrine provided general guidelines for command-line structure, but it did not clearly specify organizational roles and responsibilities for ACS supply, demand, and integrator processes. As a result, operational and combat support communities had limited understanding of who should perform which part of the ACS planning, execution, monitoring, and control process. Support was developed at the time combat support was actually being executed. Working out the ACS organizational relationships was left to the particular players who were occupying the various ACS positions, without the aid of a playbook. Under such circumstances, effectiveness can reflect the skills and experience of the players more than it reflects doctrine or policy. This lack of understanding and an ad hoc organization created problems in combat support command and control.

Many of the issues identified after JTF NA and OEF were not repeated in OIF. The long time available to plan and define relationships, coupled with the Air Force's agreement on and initial implementation of an enterprise command and control OA, greatly enhanced the command and control of combat support during OIF. ACS doctrine was in review, and some changes had been incorporated. Roles and responsibilities were tied to specific organizations. Many standing organizations used during OEF were still in place, and the leadership had recent combat experience. Individuals and organizations were better prepared to carry out their responsibilities, and the command structure was well defined. AFCENT acted as the supported command, and the rest of the Air Force supported AFCENT.

Bibliography

ACS CONOPS—*See* U.S. Air Force, 2007b.

AFDD 1—*See* Secretary of the Air Force, 2003.

AFDD 1-1—*See* Secretary of the Air Force, 2006a.

AFDD 2—*See* Secretary of the Air Force, 2007a.

AFDD 2-4—*See* Secretary of the Air Force, 2005a.

AFDD 2-8—*See* Secretary of the Air Force, 2007b.

AFI 10-201—*See* Secretary of the Air Force, 2006b.

AFI 10-244—*See* Secretary of the Air Force, 2005(2010).

AFI 10-401—*See* Secretary of the Air Force, 2006(2010).

AFI 10-403—*See* Secretary of the Air Force, 2008.

AFI 13-1AOC—*See* Secretary of the Air Force, 2005b.

AFPD 10-2—*See* Secretary of the Air Force, 2006c.

AFPD 10-4—*See* Secretary of the Air Force, 2009.

Amouzegar, Mahyar A., Ronald G. McGarvey, Robert S. Tripp, Louis Luangkesorn, Thomas Lang, and Charles Robert Roll Jr., *Evaluation of Options for Overseas Combat Support Basing*, Santa Monica, Calif.: RAND Corporation, MG-421-AF, 2006. As of May 11, 2011:
http://www.rand.org/pubs/monographs/MG421.html

Amouzegar, Mahyar A., Robert S. Tripp, Ronald G. McGarvey, Edward W. Chan, and Charles Robert Roll Jr., *Supporting Air and Space Expeditionary Forces: Analysis of Combat Support Basing Options*, Santa Monica, Calif.: RAND Corporation, MG-261-AF, 2004. As of May 11, 2011:
http://www.rand.org/pubs/monographs/MG261.html

Beer, Stafford, *Decision and Control: The Meaning of Operational Research and Management Cybernetics*, New York: Wiley, 1966.

Boyd, John R., "A Discourse on Winning and Losing," Maxwell AFB, Ala.: Air University Library, Document M-U4397, unpublished collection of briefing slides, August 1987.

Caterpillar Logistics Services, "AFGLSC Visioning Conference," briefing, Wright-Patterson AFB, Ohio, June 17, 2009.

———, "This Is Cat Logistics," fact sheet, 2011. As of May 11, 2011: http://logistics.cat.com/cda/files/1328963/7/Cat_Logistics_Fact_Sheet_2009_09.pdf

Chairman of the Joint Chiefs of Staff, *Joint Reporting Structure General Instructions*, Manual 3150.01, June 30, 1999a.

———, *Universal Joint Task List*, Manual 3500.04B, Version 4.0, October 1, 1999b. As of May 11, 2011: http://www.dtic.mil/doctrine/jel/cjcsd/cjcsm/m3500_4b.pdf

———, *JOPES Planning Formats and Guidance*, Vol. 2, Manual 3122.03B, February 28, 2006.

———, Logistics Supplement to the Joint Strategic Capabilities Plan, Instruction 3110.03C, May 9, 2007a, not available to the general public.

———, *Joint Training Manual for the Armed Forces of the United States*, Manual 3500.03B, August 31, 2007b.

———, *Joint Strategic Capabilities Plan (JSCP)*, Instruction 3110.01G, March 1, 2008a, not available to the general public.

———, *Management and Review of Campaign and Contingency Plans*, Instruction 3141.01D, April 24, 2008b.

———, *The Joint Training System: A Primer for Senior Leaders*, Guide 3501, July 31, 2008c. As of May 11, 2011: http://www.dtic.mil/doctrine/doctrine/other/g3501.pdf

———, *2009–2010 Chairman's Joint Training Guidance*, Notice 3500.01, September 8, 2008d.

CJCSG 3501—*See* Chairman of the Joint Chiefs of Staff, 2008c.

CJCSI 3110.01G—*See* Chairman of the Joint Chiefs of Staff, 2008a.

CJCSI 3110.03C—*See* Chairman of the Joint Chiefs of Staff, 2007.

CJCSM 3122.03B—*See* Chairman of the Joint Chiefs of Staff, 2006.

CJCSM 3500.04B—*See* Chairman of the Joint Chiefs of Staff, 1999b.

CJCSN 3500-01—*See* Chairman of the Joint Chiefs of Staff, 2008.

Combat Support C2: A New Vision for Global Support, Air Force Journal of Logistics, Special Edition, Vol. XXVII, No. 2, Summer 2003.

DoD—*See* U.S. Department of Defense.

DODD 7730.65—*See* U.S. Department of Defense, 2002(2007).

DRRS CONOPS—*See* U.S. Department of Defense, 2009.

Feinberg, Amatzia, Eric Peltz, James A. Leftwich, Robert S. Tripp, Mahyar A. Amouzegar, Russell Grunch, John G. Drew, Tom LaTourrette, and Charles Robert Roll Jr., *Supporting Expeditionary Aerospace Forces: Lessons from the Air War over Serbia*, Santa Monica, Calif.: RAND Corporation, 2002, not available to the general public.

Feinberg, Amatzia, Hyman L. Shulman, Louis W. Miller, and Robert S. Tripp, *Supporting Expeditionary Aerospace Forces: Expanded Analysis of LANTIRN Options*, Santa Monica, Calif.: RAND Corporation, MR-1225-AF, 2001. As of May 11, 2011:
http://www.rand.org/pubs/monograph_reports/MR1225.html

Gabreski, Maj Gen Terry L., U.S. Air Force; James A. Leftwich; Col (ret.) Robert Tripp, U.S. Air Force; C. Robert Roll Jr.; and Maj Cauley von Hoffman, U.S. Air Force, "Command and Control Doctrine for Combat Support: Strategic- and Operational-Level Concepts for Supporting the Air and Space Expeditionary Force," *Air and Space Power Journal*, Spring 2003. As of May 11, 2011:
http://www.au.af.mil/au/cadre/aspj/airchronicles/apj/apj03/spr03/gabreski.html

Galway, Lionel A., Robert S. Tripp, Timothy L. Ramey, and John G. Drew, *Supporting Expeditionary Aerospace Forces: New Agile Combat Support Postures*, Santa Monica, Calif.: RAND Corporation, MR-1075-AF, 2000. As of December 5, 2011:
http://www.rand.org/pubs/monograph_reports/MR1075.html

GDF—*See* U.S. Department of Defense, 2008a.

GEF—*See* U.S. Department of Defense, 2007.

GFMIG—*See* U.S. Department of Defense, 2008b.

Hillestad, Richard, *Dyna-METRIC: Dynamic Multi-Echelon Technique for Recoverable Item Control*, Santa Monica, Calif.: RAND Corporation, R-2785-AF, 1982. As of May 11, 2011:
http://www.rand.org/pubs/reports/R2785.html

"IBM Integrated Supply Chain (ISC) Transformation Overview: Building a Smarter Supply Chain," briefing, Wright-Patterson AFB, Ohio, June 17, 2009.

Institute for Defense Analyses, *Independent Review of DoD's Readiness Reporting System*, Alexandria, Va., 2000. As of May 16, 2011:
http://handle.dtic.mil/100.2/ADA406574

Joint Chiefs of Staff, *Joint Task Force Planning Guidance and Procedures*, Joint Publication 5-00.2, January 13, 1999.

————, *Joint Tactics, Techniques, and Procedures for Joint Theater Distribution*, Joint Publication 4-01.4, August 22, 2000.

————, *Unified Action Armed Forces (UNAAF)*, Joint Publication 0-2, July 10, 2001.

————, *The National Military Strategy of the United States of America: A Strategy for Today, a Vision for Tomorrow*, Washington, D.C., 2004. As of May 13, 2011: http://purl.access.gpo.gov/GPO/LPS59035

————, *Joint Operations*, Joint Publication 3-0, September 17, 2006, incorporating change 1, February 13, 2008.

————, *Joint Operation Planning*, Joint Publication 5-0, December 26, 2006. As of May 11, 2011: http://www.dtic.mil/doctrine/new_pubs/jp5_0.pdf

————, *Deployment and Redeployment Operations*, Joint Publication 3-35, May 7, 2007. As of May 11, 2011: http://www.dtic.mil/doctrine/new_pubs/jp3_35.pdf

————, *Doctrine for the Armed Forces of the United States*, Joint Publication 1, May 2, 2007, incorporating change 1, March 20, 2009. As of May 23, 2011: http://www.dtic.mil/doctrine/new_pubs/jp1.pdf

————, *Joint Logistics*, Joint Publication 4-0, July 18, 2008. As of May 11, 2011: http://www.dtic.mil/doctrine/new_pubs/jp4_0.pdf

————, *Command and Control for Joint Operations*, Joint Publication 3-30, January 12, 2010. As of May 23, 2011: http://www.dtic.mil/doctrine/new_pubs/jp3_30.pdf

JP 3-0—*See* Joint Chiefs of Staff, 2006 (2008).

JP 3-30—*See* Joint Chiefs of Staff, 2010.

JP 3-35—*See* Joint Chiefs of Staff, 2007.

JP 4-0—*See* Joint Chiefs of Staff, 2008.

JP 4-01.4—*See* Joint Chiefs of Staff, 2000.

JP 5-0—*See* Joint Chiefs of Staff, 2006.

JP 5-00.2—*See* Joint Chiefs of Staff, 1999.

Kent, Glenn A., *A Framework for Defense Planning*, Santa Monica, Calif.: RAND Corporation, R-3721-AF/OSD, 1989. As of May 11, 2011: http://www.rand.org/pubs/reports/R3721.html

Leftwich, James A., Robert S. Tripp, Amanda B. Geller, Patrick Mills, Tom LaTourrette, Charles Robert Roll Jr., Cauley von Hoffman, and David Johansen, *Supporting Expeditionary Aerospace Forces: An Operational Architecture for Combat Support Execution Planning and Control*, Santa Monica, Calif.: RAND Corporation, MR-1536-AF, 2002. As of May 11, 2011:
http://www.rand.org/pubs/monograph_reports/MR1536.html

Lewis, Leslie, James A. Coggin, and Charles Robert Roll Jr., *The United States Special Operations Command Resource Management Process: An Application of the Strategy-to-Tasks Framework*, Santa Monica, Calif.: RAND Corporation, MR-445-A/SOCOM, 1994. As of May 11, 2011:
http://www.rand.org/pubs/monograph_reports/MR445.html

Lynch, Kristin F., John G. Drew, Amy L. Maletic, Robert S. Tripp, Ricardo Sanchez, William A. Williams, Brent Thomas, and Max Woodworth, *A Strategic Assessment of Component Numbered Air Force (C-NAF) Force Postures*, Santa Monica, Calif.: RAND Corporation, 2010, not available to the general public.

Lynch, Kristin F., John G. Drew, Robert S. Tripp, and Charles Robert Roll Jr., *Supporting Air and Space Expeditionary Forces: Lessons from Operation Iraqi Freedom*, Santa Monica, Calif.: RAND Corporation, MG-193-AF, 2005. As of May 11, 2011:
http://www.rand.org/pubs/monographs/MG193.html

Lynch, Kristin F., and William A. Williams, *Combat Support Execution Planning and Control: An Assessment of Initial Implementations in Air Force Exercises*, Santa Monica, Calif.: RAND Corporation, TR-356-AF, 2009. As of May 11, 2011:
http://www.rand.org/pubs/technical_reports/TR356.html

McGarvey, Ronald G., James M. Masters, Louis Luangkesorn, Stephen Sheehy, John G. Drew, Robert Kerchner, Ben D. Van Roo, and Charles Robert Roll Jr., *Supporting Air and Space Expeditionary Forces: Analysis of CONUS Centralized Intermediate Repair Facilities*, Santa Monica, Calif.: RAND Corporation, MG-418-AF, 2008. As of May 11, 2011:
http://www.rand.org/pubs/monographs/MG418.html

McGarvey, Ronald G., Robert S. Tripp, Rachel Rue, Thomas Lang, Jerry M. Sollinger, Whitney A. Conner, and Louis Luangkesorn, *Global Combat Support Basing: Robust Prepositioning Strategies for Air Force War Reserve Materiel*, Santa Monica, Calif.: RAND Corporation, MG-902-AF, 2010. As of May 11, 2011
http://www.rand.org/pubs/monographs/MG902.html

Mills, Patrick, Ken Evers, Donna Kinlin, and Robert S. Tripp, *Supporting Air and Space Expeditionary Forces: Expanded Operational Architecture for Combat Support Execution Planning and Control*, Santa Monica, Calif.: RAND Corporation, MG-316-AF, 2006. As of May 11, 2011:
http://www.rand.org/pubs/monographs/MG316.html

Niblack, Preston, Thomas S. Szayna, and John Bordeaux, *Increasing the Availability and Effectiveness of Non-U.S. Forces for Peace Operations*, Santa Monica, Calif.: RAND Corporation, MR-701-OSD, 1996.

Office of the President of the United States of America, *The National Security Strategy of the United States of America*, March 2006.

PAD 06-09—*See* U.S. Air Force, 2006b.

PAD 07-13—*See* U.S. Air Force, 2008.

Peltz, Eric, Hyman L. Shulman, Robert S. Tripp, Timothy L. Ramey, and John G. Drew, *Supporting Expeditionary Aerospace Forces: An Analysis of F-15 Avionics Options*, Santa Monica, Calif.: RAND Corporation, MR-1174-AF, 2000. As of May 11, 2011:
http://www.rand.org/pubs/monograph_reports/MR1174.html

Public Law 105-261, Strom Thurmond National Defense Authorization Act for Fiscal Year 1999, October 17, 1998. As of May 13, 2011:
http://www.dod.gov/dodgc/olc/docs/1999NDAA.pdf

Schrader, John Y., Leslie Lewis, William Schwabe, C. Robert Roll Jr., and Ralph Suarez, *USFK Strategy-to-Task Resource Management: A Framework for Resource Decisionmaking*, Santa Monica, Calif.: RAND Corporation, MR-654-USFK, 1996. As of May 11, 2011:
http://www.rand.org/pubs/monograph_reports/MR654.html

Seborg, Dale E., Thomas F. Edgar, and Duncan A. Mellichamp, *Process Dynamics and Control*, New York: Wiley, 1989.

Secretary of the Air Force, *Air Force Basic Doctrine*, Air Force Doctrine Document 1, November 17, 2003. As of May 11, 2011:
http://www.dtic.mil/doctrine/jel/service_pubs/afdd1.pdf

———, *Combat Support*, Air Force Doctrine Document 2-4, March 23, 2005a. As of May 11, 2011:
http://www.dtic.mil/doctrine/jel/service_pubs/afdd2_4.pdf

———, *Operational Procedures: Air and Space Operations Center*, Air Force Instruction 13-1AOC, Vol. 3, August 1, 2005b. As of May 11, 2011:
http://www.af.mil/shared/media/epubs/AFI13-1AOCV3.pdf

———, *Reporting Status of Aerospace Expeditionary Forces*, Air Force Instruction 10-244, September 12, 2005, incorporating through change 3, September 27, 2010. As of May 11, 2011:
http://www.e-publishing.af.mil/shared/media/epubs/afi10-244.pdf

———, *Leadership and Force Development*, Air Force Doctrine Document 1-1, February 18, 2006a. As of May 11, 2011:
http://www.dtic.mil/doctrine/jel/service_pubs/afdd1_1.pdf

————, *Status of Resources and Training System*, Air Force Instruction 10-201, April 13, 2006b. As of May 11, 2011:
http://www.af.mil/shared/media/epubs/AFI10-201.pdf

————, *Readiness*, Air Force Policy Directive 10-2, October 30, 2006c.

————, *Air Force Operations Planning and Execution*, Air Force Instruction 10-401, December 7, 2006, incorporating through change 3, July 21, 2010. As of May 11, 2011:
http://www.af.mil/shared/media/epubs/AFI10-401.pdf

————, *Operations and Organization*, Air Force Doctrine Document 2, April 3, 2007a. As of May 11, 2011:
http://www.dtic.mil/doctrine/jel/service_pubs/afdd2.pdf

————, *Command and Control*, Air Force Doctrine Document 2-8, June 1, 2007b. This has been superseded by Air Force Doctrine Document 6-0.

————, *U.S. Air Force Agile Combat Support CONOPS*, November 15, 2007c.

————, *Deployment Planning and Execution*, Air Force Instruction 10-403, January 13, 2008. As of May 11, 2011:
http://www.af.mil/shared/media/epubs/AFI10-403.pdf

————, *Operations Planning: Air and Space Expeditionary Force (AEF)*, Air Force Policy Directive 10-4, April 30, 2009. As of May 11, 2011:
http://www.e-publishing.af.mil/shared/media/epubs/AFPD10-4.pdf

Sherbrooke, Craig C., *METRIC: A Multi-Echelon Technique for Recoverable Item Control*, Santa Monica, Calif.: RAND Corporation, RM-5078-PR, 1966. As of May 11, 2011:
http://www.rand.org/pubs/research_memoranda/RM5078.html

Snyder, Don, and Patrick Mills, *Supporting Air and Space Expeditionary Forces: A Methodology for Determining Air Force Deployment Requirements*, Santa Monica, Calif.: RAND Corporation, MG-176-AF, 2004. As of May 11, 2011:
http://www.rand.org/pubs/monographs/MG176.html

Thaler, David E., *Strategies to Tasks: A Framework for Linking Means and Ends*, Santa Monica, Calif.: RAND Corporation, MR-300-AF, 1993. As of May 11, 2011:
http://www.rand.org/pubs/monograph_reports/MR300.html

"Toyota Motor Saves, North American Parts Operations," briefing, Wright-Patterson AFB, Ohio, June 17, 2009.

Tripp, Robert S., Lionel A. Galway, Timothy L. Ramey, Mahyar A. Amouzegar, and Eric Peltz, *Supporting Expeditionary Aerospace Forces: A Concept for Evolving the Agile Combat Support/Mobility System of the Future*, Santa Monica, Calif.: RAND Corporation, MR-1179-AF, 2000. As of May 11, 2011:
http://www.rand.org/pubs/monograph_reports/MR1179.html

Tripp, Robert S., Lionel A. Galway, Paul Killingsworth, Eric Peltz, Timothy L. Ramey, and John G. Drew, *Supporting Expeditionary Aerospace Forces: An Integrated Strategic Agile Combat Support Planning Framework*, Santa Monica, Calif.: RAND Corporation, MR-1056-AF, 1999. As of May 11, 2011:
http://www.rand.org/pubs/monograph_reports/MR1056.html

Tripp, Robert S., Kristin F. Lynch, John G. Drew, and Edward W. Chan, *Supporting Air and Space Expeditionary Forces: Lessons from Operation Enduring Freedom*, Santa Monica, Calif.: RAND Corporation, MR-1819-AF, 2004. As of May 11, 2011:
http://www.rand.org/pubs/monograph_reports/MR1819.html

Tripp, Robert S., Kristin F. Lynch, Charles Robert Roll Jr., John G. Drew, and Patrick Mills, *A Framework for Enhancing Airlift Planning and Execution Capabilities Within the Joint Expeditionary Movement System*, Santa Monica, Calif.: RAND Corporation, MG-377-AF, 2006. As of May 11, 2011:
http://www.rand.org/pubs/monographs/MG377.html

Tripp, Lt Col Robert S., and Capt Larry B. Rainey, Air Force Logistics Command, Deputy Chief of Staff, Plans and Programs, Directorate of Special Projects, "A Cybernetic Approach for the Design and Development of Management Information and Control Systems (MICS): An Illustration Within the Air Force Logistics Command," *Cybernetica*, Vol. XXVI, No. 4, 1983, pp. 282–305.

———, "Cybernetics: A Theoretical Foundation for Developing Logistics Information and Control Systems," *Logistics Spectrum*, Summer 1985, pp. 32–38.

Tripp, Robert S., Larry B. Rainey, and John M. Pearson, "The Use of Cybernetics in Organizational Design and Development: An Illustration Within the Air Force Logistics Command," *Cybernetics and Systems: An International Journal*, Vol. 14, No. 2–4, 1983, pp. 293–314.

Tripp, Robert S., William A. Williams, Kristin F. Lynch, John G. Drew, Dahlia S. Lichter, and Laura H. Baldwin, *A Strategic Analysis of Air and Space Operations Center Force Posture Options*, Santa Monica, Calif.: RAND Corporation, 2008, not available to the general public.

U.S. Air Force, *Air Force Forces Command and Control Enabling Concept*, change 2, May 25, 2006a.

———, *Implementation of the Chief of Staff of the Air Force Direction to Establish an Air Force Component Organization*, Program Action Directive 06-09, November 7, 2006b.

———, *Agile Combat Support Concept of Operations (ACS CONOPS)*, November 15, 2007.

———, *Implementation of the Chief of Staff of the Air Force Direction to Transform and Consolidate Headquarters Management Functions*, Program Action Directive 07-13, January 25, 2008.

———, *U.S. Air Force Defense Readiness Reporting System (DRRS) Interim Guidance*, February 23, 2009a.

———, *U.S. Air Force Universal Task List*, March 4, 2009b.

———, *Annual Planning and Programming Guidance (APPG) FY 11–15*, May 2009c, not available to the general public.

———, *Implementation of the Chief of Staff of the Air Force Direction to Restructure Command and Control of the Component Numbered Air Forces*, Program Action Directive 10-02, draft, November 9, 2009d.

———, *Agile Combat Support Core Function Master Plan 2010*, draft, January 19, 2010a.

———, *Command and Control Service Core Function Master Plan, 2010–2030*, draft, January 19, 2010b.

U.S. Code, Title 10, Armed Forces, Subtitle A, General Military Law, Part I, Organization and General Military Powers, Chapter 2, Department of Defense, Section 117, Readiness reporting system: establishment; reporting to congressional committees.

U.S. Department of Defense, *Department of Defense Readiness Reporting System (DRRS)*, Directive 7730.65, June 3, 2002, certified current as of April 23, 2007. As of May 11, 2011:
http://www.dtic.mil/whs/directives/corres/pdf/773065p.pdf

———, *Guidance for Employment of the Force (GEF)*, draft, August 31, 2007, not available to the general public.

———, *Guidance for Development of the Force (GDF)*, April 2008a, not available to the general public.

———, *Global Force Management Implementation Guidance (GFMIG) FY08–09*, June 4, 2008b, not available to the general public.

———, *Defense Readiness Reporting System Concept of Operations*, Version 3.0, January 22, 2009. As of May 11, 2011:
http://certification.drrs.org/files/0000/0034/DRRS_CONOPS_2009.pdf

———, *DoD Dictionary of Military and Associated Terms*, Joint Publication 1-02, November 8, 2010, as amended through October 15, 2011. As of December 5, 2011:
http://www.dtic.mil/doctrine/dod_dictionary/

Wickman, Allen, deputy chief of staff, Operations, Plans, and Requirements, Command and Control Operations, Employment and Force Development, and Col Patricia Battles, deputy chief of staff, Installations and Mission Support, Global Combat Support, "C-NAF AFFOR C2 (Draft/Predecisional)," Air Force Command and Control General Officer Steering Group Meeting, Andrews AFB, Md., November 17, 2009.

Zettler, Lt Gen Michael E., U.S. Air Force, "The New Vision," *Air Force Journal of Logistics*, Vol. 27, No. 2, Summer 2003, pp. 4–6.